RhomboSTEAM

K-2 adventures in geometry

for

2D Shapes

SECOND GRADE

STEAMBRIDGE PRESS

Charlotte, North Carolina
Portland, Oregon

Published by STEAMBridge Press
2901 North Davidson Street
Charlotte, NC 28201

ISBN 978-1-942357-27-8

Photo permissions and authors, illustrators, designers, photographers available for download at the URL.

http://www.streambridgepress.com

STEMWORKS
CERTIFIED
READY-TO-SCALE

STEMWORKS EXEMPLARY PROGRAM STATUS

Ten80 Education and its programs are included in the STEMWorks Database by Change the Equation, a non-partisan, CEO-led initiative to connect and align efforts to improve STEM learning. Ten80 Education programs met the highest standards for excellence and exemplary STEM project based programming through rigorous examination by WestEd, an independent nonprofit research organization.

RECOMMENDED BY REAL TEACHERS

We are thoroughly enjoying our STEAM lessons. The program is excellent! My students have loved making the Rhombis and getting to build with shapes and tape. It has been a big hit! Thanks for everything!
Alicia Stenard
Mater Christi in Albany School District

"We need more projects like this that carry on for a longer number of weeks. My teachers used scope and sequence suggestions, and the program ran after school for a semester. We could have worked on this a lot longer."
Karen James
Druid Hills Elementary in Charlotte, NC

"I'm excited for our students to get their hands on real STEAM projects. There isn't much out there for Kindergarten and First Grade."
Sandy Mettler
Fort Worth Schools in Dallas, TX

RHOMBI'S ADVENTURES IN 2D
EDITOR'S COMMENTS

To the Educators using RhomboSTEAM,

First of all, thank you for taking the leap. Project-based learning requires a commitment on the part of the teacher. It can be messy. It can mean more planning and more time spent gathering odds and ends that make a project work. The time and energy you spend developing imaginative responses and ability to problem solve will pay off tenfold.

There is significant evidence in our own research that indicates a need for memorization and rote learning IN ADDITION TO LONG-RANGE PROJECTS and innovative problem-solving. A decent project uses the challenge as a springboard for planning skills-based lessons. A great project urges kids to seek skills and knowledge that you can offer because they want to succeed in solving the problem you've posed. Quick recall of facts is necessary for innovative thinking. Kids don't need to stop and look up simple facts or count on teddy bears when we want them to focus on collecting data and making reasonable decisions.

Projects need outcomes. They need number-driven goals that qualify success in the challenges. Let kids get creative and try things that will fail. Did I say to let them fail? Let them fail. Innovators fail all the time. Some of our greatest inventions are the result of absolute and profound failures. Learning to use the failure as a tool rather than an emotional trial is where students are headed. It takes a few falls before learning to ride a bike. They will make a few bouncy balls on the road to a great slimy wall-walker.

The question arises as to why we chose to call students "designers" rather than "engineers" or "scientists." Designers incorporate a wide variety of communication skills as well as industry expertise. Because communication of ideas among students is integral to the success of any project-based curriculum, the students become Rhombi's designers.

Ten80 celebrates the amazing lessons we learn from you. When you develop a new technique or unique approach to the material, let us hear from you. The resource site grows stronger as teachers share their ideas with each other.

Sincerely,

Jeannie Ruiz
President of Ten80 Elementary

DESIGNERS, WILL YOU HELP ME?

Designers collect, analyze, and synthesize data guided by specifications.

Designers make clear and concise recommendations using drawings,verbal descriptions, and math models.

The Industrial Designers Society of America's definition of industrial design describes design as "the professional service of creating and developing concepts and specifications that optimize the function, value and appearance of products and systems for the mutual benefit of both user and manufacturer."

Rhombi asks her friends to offer assistance in design. Students collect, analyze and make data-based decisions. They must share their concepts verbally, using drawings, and through models (scale and math).

RHOMBI'S ADVENTURES IN 2D
FIND YOUR RESOURCES

ACTIVITY		HANDOUT	READ ALOUD	
ROUND AND ROUND *PAGE 18*				
HARVEST SHAPES	23	136	LUNA AND SOL'S BIG ADVENTURE	8
PATTERNS	24	137	ROVER, ROVER COME ON OVER	10
WHEELS	25	138	WHEELS ON THE GROUND	12
QUILTS	26	139	HANGING WITH CIRCLES	14
CIRCLES SHAPE WALK	27	140	RADIUS AND PI	16
LOOKING FOR SQUARES *PAGE 40*				
GROWING THINGS	45	141	SQUARE FOOT HARVEST	20
QUADRILATERALS	46	142	P IS FOR PARALLEL	22
SQUARES SHAPE WALK	47	143	SHOPPING FOR SHAPES	24
QUILT PATTERNS	48	144	STORIES IN THE SQUARE	26
PAPER SQUARES	49	145	PAPER FORESTS	28
TASTY TRIANGLE *PAGE 62*				
POLYGONS	67	146	RICKETY RACKETY	32
PRESENTATIONS	68	147	CITY OF THREE ANGLES	34
GROWING	69	148	TRAVELING TRIANGLE	36
TRIANGLES IN ART	70	149	FOX'S FOREST	38
WHAT'S YOUR ANGLE	71	150	TRI, TRI AGAIN	40
FLYING FIVES *PAGE 86*				
STAKE OUT THE PENTAGON	91	151	WE'RE HUNTING FOR PENTAGONS	44
TOOLS THROUGH TIME	92	152	TOOL TIME	46
MATERIALS	93	153	CREATIVE DESIGN	48
STORY IN 5 BOXES	94	154	DESIGN TME	50
WALK THE PENTAGON	95	155	THE PENTAGON, USA	52
ON THE MOVE *PAGE 108*				

INTRO TO STEAM

4 UNITS 1-5: QUICK OVERVIEW
6 MATH MODELING IN K2 SETTINGS
7 PROBLEM-BASED LEARNING
8 GREAT S.T.E.A.M. TEACHER CHECKLIST
9 CONFIDENT S.T.E.A.M. STUDENTS
10 ADULT MISCONCEPTIONS
12 DEFINING S.T.E.A.M.
13 BUILDING S.T.E.A.M. IN CLASSROOMS
15 GEOMETRY IS THE THREAD
16 LONG RANGE PROJECTS
17 USING THE MATERIALS

APPENDIX

123 NETS
125 POLYGON TRACING PAGES
134 POLYGON BACKGROUND
135 STEMVESTIGATIONS HANDOUTS
156 MINDBUGS IN MEASUREMENT
163 VOCABULARY EXTENSIONS
164 PRE AND POST ASSESSMENTS
173 PLANNING PAGES
175 NEXT GEN SCIENCE STANDARDS
176 COMMON CORE CHECKLISTS
183 S.T.E.A.M. WORD WALL
189 REFERENCES AND WORKS CITED

RHOMBI'S ADVENTURES IN 2D
UNITS OVERVIEW

ROUND AND ROUND (Play with "Circle, Circle, Rhombus") 18

Designers will help Rhombi devise a wheeled cart for transporting a telescope.

SCIENCE: Chart the phases of the moon.

TECHNOLOGY: Take virtual field trips to the moon and beaches around the world.

ENGINEERING: Create a recycled tray that carries 6 oz for 12 inches with the least impression in sand.

ART: Use clay circles to create sneaker prints.

MATHEMATICS: Measure to find radius and make "our circles."

LOOKING FOR SQUARES (Play with "How Many Edges.") 40

Designers will help Rhombi create a paper tool that helps to locate squares.

SCIENCE: "Harvest" shapes found in the lunchroom. Use tallies to track and count on a chart.

TECHNOLOGY: Mobile digital walk gathers shapes found in the school for digital presentations.

ENGINEERING: Look for squares in the immediate environment and connect each square with its use.

ART: Create paper quilts using quadrlaterals. Incorporate measurement and counting.

MATHEMATICS: Learn to fold a square as the basis for origami projects and assess papers for folding.

RHOMBI'S ADVENTURES IN 2D
UNITS OVERVIEW

TASTY TRIANGLE (Play with "Shape Mimes.") 62

Designers will help Rhombi with a mousy bridge supported by triangles.

SCIENCE: Determine the strongest materials for triangle construction.

TECHNOLOGY: Use apps and Internet sites to interact with shapes on a virtual geoboard and more.

ENGINEERING: Investigate structures with platforms. Build square and triangle based creations.

ART: Play with primary and secondary colors using the triangle.

MATHEMATICS: Reshape the world with polygons that cover images from magazines.

FLYING FIVES (Play with "Dot to Dot") 86

Designers will help Rhombi create a pentagonal parachute for a camping gear drop.

SCIENCE: What's in a pentagon? Take a closer look at the ground inside a regular pentagon.

TECHNOLOGY: Optimize and innovate by designing a tool to improve life in the classroom.

ENGINEERING: Make and use pentagon patterns.

ART: Discover the fun of informational text within a comic strip. Talk about pentagons!

MATHEMATICS: Walk the walls of the pentagon and build its scale model (well, the walls, anyway).

ON THE MOVE 86

Designers will take part in an authentic assessment as they troubleshoot a camping trip.

RHOMBI'S ADVENTURES IN 2D
MATH MODELING IN K-2 SETTINGS

DEFINING THE MATH MODEL

Mathematics can be used to "model", or represent, how the real world works. A model is not the real thing, but it should be accurate enough to be useful. Math models connect abstract numbers to real experience. Math models save time, supplies, and aggravation. For example, Rhombi needs to build a full-scale cart. Since the class cannot build the full sized vehicle, it will develop a small scale version and extrapolate from there. They use their background knowledge to begin the math model.

MOVE FROM EXPERIENCE TO SYMBOLS

The study of science and math started in an effort to understand actions that could be observed directly. In order to make science and math a useful performing art, instruction must start with real experience and gradually build toward an abstract representation. Doing things in the reverse chronological order produces nearly useless results.

KNOW WHERE THE LESSON IS HEADED

Education separates experience from the learning process. We could learn a few things from our own history. Math development progressed along its slow, continuous track: caveman to calculus. Here, in the 21st century, we expect second graders to operate at abstract levels using a symbolic language that took mathematicians thousands of years to develop. Educators are asked to start with the algorithm –the abstraction- and work backwards to the concrete. We wonder why math is such a dreaded subject. Neither the teacher nor the student knows where the symbols are supposed to take them.

STEAM USES NATURAL EXTENSIONS

Abstract models of STEAM phenomena should grow as a natural extension of what is felt by direct experience. The abstraction must be an immediate and obvious representation of what students already know at every level. Drive a radio controlled car. Build the shapes as you memorize them. Investigate, collect data and analyze results. From reality to abstraction, STEAM learning needs more teachers who understand that doing more means going further, faster.

DEFINE YOUR MATHEMATICAL MODEL

Identify the problem, define the terms in your problem, & draw diagrams where appropriate. Begin with a simple model, stating assumptions that you make as you focus on particular aspects of the phenomenon. Identify important variables and constants. Determine how they relate to each other. Develop the equation(s) that express relationships between variables & constants.

RHOMBI'S ADVENTURES IN 2D
PROJECT BASED LEARNING

Learning shapes is great. Doing short-term interactive projects that involve shapes is even better. Turning those projects into inquiry-based challenges takes things a step further. Involving students in long-range projects that require multiple steps, depth of knowledge, and true integration of science and math is the goal. The National Science Foundation funded Innovative Technology Experiences for Teachers (ITEST) research that developed a list of criteria for excellence in project based learning.

CRITERIA FOR EXCELLENCE IN PROJECT BASED LEARNING

- Focus on activities that motivate students to develop interest and skills in STEM careers.
- Provide intensive engineering experience of a long duration (academic year and/or summer proportional to the nature of the program).
- Use technlogy tools and hands-on learning.
- Involve industry mentors, scientists, and engineers.
- Incorporate students' self-regulated learning within specific engineering design experiences.
- Offer innovative, high interest STEM activities.
- Involve participants in the designing and building of artifacts.
- Use scientific integration and/or the engineering design process.
- Engage students in project-based and/or problem-based learning.
- Develop participants' technical skills (as appropriate to the project).
- Make explicit connections between (STEM) academics and project.
- Build in success experiences and opportunities to learn from failure.
- Explore technical careers from technicians to engineers and engineering management.

RHOMBI'S ADVENTURES IN 2D
GREAT S.T.E.A.M. TEACHER CHECKLIST

1. **DO MORE THAN IMPART KNOWLEDGE** Children will teach themselves if offered the opportunities to explore and problem solve. Teach by posing questions. Teach by developing curiosity. Teach by guiding learners with interactive experiences. Teach to learn. Learn as much as you teach. Adults often take so much control of the teaching process that children resort to mimicry as a way of pleasing their mentors. Every child learns at a unique pace and with a unique set of circumstances; therefore each child requires a unique path. If educators view learning as a problem-solving process, learning becomes an individual journey.

2. **DISLODGE THE MISCONCEPTIONS** Every child enters school with a unique set of skills and preconceptions about learning. Every child has already formed opinions based on observation, mimicry, and the outcome of challenges (how were successes and failures handled by adults). Some facts that lodge in a child's mind are erroneous. Other ideas are keen observations of the natural and technological world. Facilitate learning. Increase knowledge. Guide inquiry. Open minds to the possibilities. Offer multiple opportunities to learn and approaches to problem solving as a way of addressing misconceptions.

3. **USE WHOLE GROUP SPARINGLY** There are rare teachers with the ability to engage learners as a group. This interaction has a fast-paced Q&A quality and –generally- involves a points system or prizes. The up-side is that learners are involved as a group and learn courteous behavior. The down-side is that educators are rarely able to get individual feedback or offer reinforcement / stimulation. Even technologically based feedback systems are an impersonal way to discover a student's strong and weak areas.

4. **LEARN MORE WITH SMALL GROUPS** Teachers that interact in small groups with students on a regular basis know far in advance of testing exactly what the outcomes will be. Anxiety lessens for teachers and learners. Differentiation becomes possible, and students no longer "tune out" the instructor when the instructor becomes a facilitator. It's hard to ignore the teacher when she's sitting at your table asking questions and challenging assumptions. It's difficult to avoid detection if the investigation requires active participation. For the learner, there is no more "tuning out" or sleep-walking through the day.

5. **ASSESS STUDENTS REGULARLY** There are no surprises when learners take part in authentic and active assessment with reinforcement for some and mental stimulation for others. Great teachers know well ahead of time how students will perform on standardized tests.

RHOMBI'S ADVENTURES IN 2D
CONFIDENT S.T.E.A.M. STUDENTS

Children View Themselves as Scientists in the Process of Learning.
1. They look forward to doing science.
2. They demonstrate a desire to learn more.
3. They seek to collaborate and work cooperatively with their peers.
4. They are confident in doing science;
 they demonstrate a willingness to modify ideas, take risks, and display healthy skepticism.

Children Accept an "Invitation to Learn"and Readily Engage in The Exploration Process.
1. Children exhibit curiosity and ponder observations.
2. They move around selecting and using the materials they need.
3. They take the opportunity and the time to "try out" their own ideas.

Children Plan and Carry Out Investigations.
1. Children design a way to try out their ideas, not expecting to be told what to do.
2. They plan ways to verify, extend or discard ideas.
3. They carry out investigations by: handling materials, observing, measuring, and recording data.

Children Communicate Using a Variety of Methods.
1. Children express ideas in a variety of ways: journals, reporting out, drawing, graphing, charting, etc.
2. They listen, speak and write about science with parents, teachers and peers.
3. They use the language of the processes of science.
4. They communicate their level of understanding of concepts that they have developed to date.
 Children Propose Explanations and Solutions and Build a Store of Concepts.

Children offer explanations from a "store" of previous knowledge.
1. They use investigations to satisfy their own questions.
2. They sort out information and decide what is important.
3. They are willing to revise explanations as they gain new knowledge.

Children Raise Questions.
1. Children ask questions (verbally or through actions).
2. They use questions to lead them to investigations that generate further questions or ideas.
3. Children value and enjoy asking questions as an important part of science.

Children Use Observation.
1. Children observe, as opposed to just looking.
2. They see details, they detect sequences and events; they notice change, similarities and differences, etc.
3. They make connections to previously held ideas.

Children Critique Their Science Practices.
1. They use indicators to assess their own work
2. They report their strengths and weaknesses.
3. They reflect with their peers.

"Inquiry Based Science: What Does It Look Like?" Connect Magazine (published by Synergy Learning), March-April 1995, p. 13.

RHOMBI'S ADVENTURES IN 2D
ADULT MISCONCEPTIONS

These misconceptions were identified based on in-depth interviews with early childhood teachers about the key issues in early mathematics education as well as researchers' experiences in teaching early childhood students, conducting workshops with early childhood teachers (Ginsburg, Jang, Preston, VanEsselstyn & Appel, 2004; Ginsburg et al., 2006), working with them in early childhood classrooms, and engaging in informal conversations with them. The study also based descriptions of the myths on available research literature (Ginsburg, Lee & Boyd, 2008). The nine misconceptions defined by Joon Sun Lee of Hunter College, The City University of New York and Herbert P. Ginsburg of Teachers College, Columbia University are:

1. Young children are not ready for mathematics education.
2. Mathematics is for some bright kids with mathematics genes.
3. Simple numbers and shapes are enough.
4. Language and literacy are more important than mathematics.
5. Teachers should provide an enriched physical environment, step back, and let the children play.
6. Mathematics should not be taught as stand-alone subject matter.
7. Assessment in mathematics is irrelevant when it comes to young children.
8. Children learn mathematics only by interacting with concrete objects.
9. Computers are inappropriate for the teaching and learning of mathematics.

These misconceptions often interfere with understanding and interpreting the new recommendations on sound early childhood mathematics education, and become subtle (and sometimes overt) obstacles to implementing the new practices in the classrooms (Richardson, 1996).

A WORD FROM TEN80 TO TEACHERS

Become aware of your personal biases. Just because you had poor math instruction does not mean math is hard.

The language of math cannot be taught with the same symbolic approach as the language of reading and writing. Because everyone talks and practices the art of communication, teachers assume that words are the only way to demonstrate proficiency in every field including math and science.

Math and science are performing arts that must be demonstrated in performance. Math and science must use pictures and numbers depicting an actual experiment that is a test of the student's abstract model of some aspect of reality. In this way, students build their own understanding of the patterns that underlie math and science. (Ten80 2002)

RHOMBI'S ADVENTURES IN 2D
DEFINING S.T.E.A.M.

STEAM (science, technology, engineering, arts, mathematics) is just now becoming an accepted acronym among teachers and is seen as a somewhat mysterious –almost magical- fix for all the problems American schools and businesses are facing. Perhaps we should start with what STEAM is not:

- STEAM is not a conglomeration of lessons culled from five subject areas.
- STEAM is not something you can cover in a unit or two every year.
- STEAM is not a 5-day or 10-day curriculum piece.
- STEAM cannot be delivered in the traditional pencil-and-paper method by a sage on the stage standing behind the classroom lectern or seated behind a grand old wooden desk. There are no "Buehler..? Buehler..?" moments in a STEAM environment. *(reference the movie Ferris Buehler's Day Out)*

SO WHAT IS STEAM?

STEAM is a philosophy of teaching, learning and working. STEAM is more than just the sum of the words in its acronym.

In other words, do real stuff to find real answers to real problems.

STEAM IS PRETTY SIMPLE TO STEAM PROFESSIONALS

What has become a raging debate among educators seems pretty simple to the professionals that actually work in STEAM fields. Academicians will spout research from their PhD thesis and published articles in journals, etc. Ask a scientist, a mathematician, an engineer or a programmer how to prepare kids for a STEAM workplace and they'll tell you to make kids DO something.

SET UP CLASSROOMS FOR DISCOVERY

While activity centers are typical in kindergarten settings, they have all but disappeared by second grade. Bring them back. Small group interaction is optimal for RhomboSTEAM as it is for any STEAM program. Why do you want to use stations or centers in the classroom?

- Simplifies planning - plan over a week rather than day by day...
- Offers opportunities for more cohesive lessons...
- Allows teachers to hold small group sessions daily...
- Encourages teamwork...
- Simplifies classroom management... *
- STEAM increases the ability to draw connections among various concepts.

* reference Harry Wong and **The Effective Teacher**
https://www.effectiveteaching.com

RHOMBI'S ADVENTURES IN 2D
BUILDING S.T.E.A.M. CLASSROOMS

According to corestandards.org, "The Common Core State Standards provide a consistent, clear understanding of what students are expected to learn, so teachers and parents know what they need to do to help them. The standards are designed to be robust and relevant to the real world, reflecting the knowledge and skills that our young people need for success in college and careers. With American students fully prepared for the future, our communities will be best positioned to compete successfully in the global economy."

Support Common Core Standards and Learning Goals

STEAM PROJECT STATION

Support the ongoing STEAM project with a permanent station and construction materials. Recycled and upcycled materials should be available for use along with tools, paper, pencil and data collection opportunities.

ART, MUSIC, DRAMA

Encourage discussion of STEAM subjects through the arts as well as investigations involving art inventions, art history, genres, musical instruments and sound, etc. Create a "stage" where students give skits, give presentations, teach lessons and make up songs about classroom learning.

BLOCK PLAY

Keep blocks available all the time along with small scale toys at different scales. Add books about world architecture, habitats, and building projects.

NONFICTION LIBRARY

Facilitate the transition from fiction and storybooks to gleaning facts from informational text. Fill your library with fiction AND non-fiction. Encourage students to write informational text and to keep journals for science and math.

STEAM CHARACTER ROLE PLAY

Most kindergarten classrooms include an area for dress-up and pretend. How many first and second graders would still benefit from this type of role-play? Shoot for higher level characterization with lab coats, construction hats, medical scrubs, artist's smocks, etc.

RHOMBI'S ADVENTURES IN 2D
GEOMETRY IS THE THREAD

The word "geometry" comes from the Greek words for "Earth" and "measure." Geometry was first used to measure and chart the length, area and shape of land surfaces.

GEOMETRY is the single math thread that runs through Common Core from Kindergarten to 12th grade. RhomboSTEAM promotes STEAM and the Art of STEM through an authentic experience in engineering problem solving built on a platform of geometric constructions, analytical geometry, and engineering design. Math is a gateway to success in professional fields: speaking the language of mathematics fluently means that students may define their career paths by choice, not resignation.

~ Jeannie Ruiz

RhomboSTEAM Developer and Director of STEM Ten80 Foundation

THE MATH GENE
Contrary to popular belief, no one is born with a genetic predisposition toward math. There is no "math gene" in humans (Keith Devlin, The Math Gene); mathematics is a learned skill for everyone just like basketball, playing the French Horn, or ballet. Humans have brains which are already wired to enable math acquisition (Stanislas Dehaene, Number Sense 1997). Infants, chimps, dogs, and birds can recognize groups of objects up to three and decide that an object has been taken away or added. That is not "doing math," but it is a qualitative foundation on which to build understanding. Math instruction in school develops the initial qualitative sense of space, time, and quantity into the descriptive -and highly useful- language of mathematics. Children should learn all things in context.

MULTI YEAR FLUIDITY
Geometry as a foundation allows the educator to implement multi-year projects that students enhance year after year. The essence of the project is still fluid, while monthly, weekly, and daily objectives are met for the grade level.

NEXTGEN SCIENCE STANDARDS - STRUCTURE AND FUNCTION
Engineering and Design: Students should understand that "the shape and stability of structures of natural and designed objects are related to their function(s).
(K-2-ETS1-2)."

RHOMBI'S ADVENTURES IN 2D
LONG RANGE PROJECT

If we had a floating house (school, city, state) that could go anywhere on the ocean, where would we want to be?

Instead of being planted in one place, what if we could take our house (school, city, state) with us? Turtles do. No matter where the turtle goes, it's always home.

Will we stay in the same place all the time, or move with currents, weather patterns and seasons?

Build a room, house, school, city, state, or country as a floating island.
Basic elements are squares and triangles which combine (add) to form a solid. Students begin to seek skill in STEAM subjects as a way to solve problems and overcome challenges... as long as the project is interesting enough to keep their attention.

Begin investigating:

- How much milk is delivered to your school everyday?
- Where does it come from?
- What is the source of milk on the floating city-state island?
- How many people can live in a defined habitat? What do they require for survival and comfort?
- How will you move from island to land?
- What happens to structures in various kinds of weather?
- Can we stay out of the way of huge storms over the ocean?
- Do we want to move with the seasons?
- Will it move? How fast do things move?
- If we move, what will fuel the island?
- How much will all that cost, by the way?
- Where is the funding source?

and on... and on... and on...

Since we can't build full-size during the planning phase, we'd better start with scale models. The models will save time, energy, materials, cost and headache. Let's start with the basic information on geometry, polygons and structures.

Introducing RhomboSTEAM.
"Once upon a time, there was Rhombi. She loved shapes..."

RHOMBI'S ADVENTURES IN 2D
USING THE MATERIALS

ENGAGE Each activity uses a read-aloud book to jumpstart the investigation and engage students.

DISCUSS Establish mindbugs with an introductory discussion for each investigation. The MindBug is a misconception in student thinking; procedural understanding that worked in one situation but is incorrectly applied to other skills. Like a computer virus, MindBugs spread through a student's system.

Example In Science:
Children often think that air weighs nothing because they see scales measuring zero from an early age.

Example in Math:
Students line the numbers up on the right to add. This works until 5th grade when they attempt to add decimals.

EXPLAIN Each activity includes a brief background for the purpose of introducing the subject matter.

MOVE Get students up and moving to activate that extremely useful muscle memory.

INVESTIGATE

Although each EXPLORE activity and investigation stands alone, together they also build knowledge required to meet the long-range project. In Module 1, students will -ultimately- be challenged to help Rhombi prepare for a trip.

ELA Build skills in communication, writing and speaking.

UNIT 1
ROUND AND ROUND

Essential Questions

1. How would you describe the moon's shape?
2. What units should be used to measure length?
3. Can you talk about shapes using attributes?
4. How are rotation and revolution different?
5. What are you trying to accomplish in Rhombi's Challenge?

ENDURING UNDERSTANDING

Two dimensional shapes create new shapes.

"Halves" means two equal parts."

Circles and rectangles break into equal parts.

Breaking circles or rectangles into more equal parts means that the parts will be smaller.

Three objects may be put in order from longest to shortest by comparing their lengths.

The length of an object may be discussed using whole numbers.

A smaller object may be used as a measurement tool when measuring longer lengths.

Larger numbers consist of tens and ones.

Organizing information helps in discussion.

ROUND AND ROUND

Focus on circles and things that move in circles. Learn about rotation and revolution with a focus on the moon. Develop weight-carrying rovers. Dance to the lunar cycle. Explore quilt patterns and create radius art.

Students will apply basic measurement, understanding of number, place value, data collection, and data application in problem solving.

Rhombi wants to move her telescope from one location to another. Examples might include a horse or donkey cart, a wagon, or a push cart.

Watch for these extension level words:

astronomer
a person who is skilled in astronomy or who makes observations of celestial phenomena

Galileo
Galileo Galilei (1564–1642) was the first astronomer to use a telescope. He discovered 4 moons orbiting Jupiter and supported Copernicus' theory of a heliocentric universe.

revolution
the action of moving around something in a path that is similar to a circle

rotation
a complete turn around a central point

Rhombi needs a cart round wheels to haul a telescope.

MATH BACKGROUND - Circles

A circle is the set of points in a plane that are equidistant from a given point.

Radius is the distance from center. Diameter is twice the radius. From its center, the circle's angle extends to a full 360 degrees.

TECHNOLOGY CONNECTIONS

Visit the blog for links to excellent STEAM geometry activities and investigations.

- Geometry Games for Kids
- Book Lists
- Links to Apps

HISTORICAL CONNECTIONS

Natural circles like the Sun and Moon exist.

Evidence of wheeled vehicles appears from the mid-4th millennium BC in the Neolithic age.

The first recorded algorithm for Pi was developed in 250 BC by the Greek mathematician Archimedes**.

* http://en.wikipedia.org/wiki/Circle#History
** http://en.wikipedia.org/wiki/Pi#History

ART CONNECTIONS

Artists have used circles since earliest recorded history. From mandalas to crop circles, circles appear in all cultural art. Modern artists present the greatest opportunity for class experiences.

Joan Miro was a Spanish painter, sculptor, and ceramicist (1893-1983). http://joanmiro.com/

Circles in a Circle from 1923 by Vasily Kandinsky, (1866 - 1944) is an oil on canvas.
http://www.wassilykandinsky.net/

Download free mandala patterns.
http://www.free-mandala.com/

READING CONNECTIONS

Circle Unbroken
by Margot Theis Raven

So Many Circles, So Many Squares
by Tana Hoban

Pond Circle
by Betsy Franco

Pumpkin Circle: The Story of a Garden
George Levenson

ROUND AND ROUND
CURRICULUM CONNECTIONS

COMMON CORE CONNECTIONS

ELA/Literacy
W.K.7 Participate in shared research and writing projects (e.g., explore books by a favorite author and express opinions). (K-ESS2-1) add venn

Mathematics
MP.2 Reason abstractly and quantitatively. (K-ESS2-1)
MP.4 Model with mathematics. (K-ESS2-1)
K.CC.A Know number names and the count sequence. (K-ESS2-1)
K.MD.A.1 Describe measurable attributes of objects, such as length or weight. Describe several measurable attributes of a single object. (K-ESS2-1)
K.MD.B.3 Classify objects into given categories; count the number of objects in each category and sort the categories by count. (K-ESS2-1)

NEXT GEN SCI CONNECTIONS

K-2-ETS1-1. Ask questions, make observations, and gather information about a situation people want to change to define a simple problem that can be solved through the development of a new or improved object or tool.
K-2-ETS1-2. Develop a simple sketch, drawing, or physical model to illustrate how the shape of an object helps it function as needed to solve a given problem.
K-2-ETS1-3. Analyze data from tests of two objects designed to solve the same problem to compare the strengths and weaknesses of how each performs.
K-PS2-1. Plan and conduct an investigation to compare the effects of different strengths or different directions of pushes and pulls on the motion of an object.
K-PS2-2. Analyze data to determine if a design solution works as intended to change the speed or direction of an object with a push or a pull.*
1-ESS1-1. Use observations of the sun, moon, and stars to describe patterns that can be predicted.

RESOURCES FOR UNIT 1: S.T.E.A.M. ACTIVITIES FOR "ROUND AND ROUND

Rhombi Audio Download and Videos Available WINTER 2015

UNIT 1: Round and Round: Science Activity:	Document the moon's shape over one month.
UNIT 1: Round and Round: Technology Activity:	Take a virtual interactive lunar tour.
UNIT 1: Round and Round: Engineering Activity:	Design and improve Rhombi's mode of travel.
UNIT 1: Round and Round: Art & Music Activity:	Create a paper quilt featuring overlapping circles.
UNIT 1: Round and Round: Mathematics:	Use string to measure and make math art.

UNIT 1: ELA Activity: Read "Round and Round" aloud. Students illustrate the story.

BUILD STEAM WITH GREAT BOOKS

SCIENCE

So That's How the Moon Changes Shape
by Allan Fowler

Man on the Moon
by Anastasia Suen

TECHNOLOGY

Cars on Mars: Roving the Red Planet
by Alexandra Siyon

Looking Through a Telescope
by Linda Bullock

ENGINEERING

Henry Ford: Big Wheel in the Auto Industry
by Mike Venezia

What Do Wheels Do All Day?
by April Jones Prince

The ARTS

Art & Lots of Dots
by Ginett Alarcon

Eight Hands Round: A Patchwork Alphabet
by Ann Whitford Paul

MATH

When a Line Bends . . . A Shape Begins
by Rhonda Gowler **Greene**

Shapes in Math, Science and Nature: Squares, Triangles and Circles
by Bill Slavin

ELA

Ladder to the Moon
by Maya Soetoro-Ng

Papa, Please Get the Moon for Me
by Eric Carle

STEMVESTIGATIONS: MOON PHASES

SCIENCE

READ **Luna and Sol's Big Adventure** *page 8*

DISCUSS What do students already think they know about the moon? Make a list to identify "mindbugs."

EXPLAIN The moon rotates and revolves. The Earth's shadow on the moon makes its shape appear to change. The moon reflects light but does not give off its own light. The circular moon we see is actually a sphere, but we see only one side.

MOVE Rotate in place like the Earth or Moon or Sun. Revolve around a rotating student. Try rotating and revolving around the rotating student. Shine a light to simulate the sun. Draw the shapes or snap a photo to show that shadows on the students will also show full, half, crescent, new, crescent, half, full.

INVESTIGATE

Create a chart to mark the moon's shape each day for a month. Assign a quick sketch each day as student homework for the month. Each team of students should be required to demonstrate the Sun's light shining on the Moon. Then each team should be required to demonstrate Earth shadows on the Moon. (Assign one phase of the moon to each team, or have each team demonstrate all 4 major moon phases.)

ELA Have teams create plays to demonstrate phases.

HANDOUT PAGE 137

TECHNOLOGY

READE

READ **Rover, Rover, Come on Over** *page 10*

DISCUSS Rovers are vehicles handled by remote drivers and capable of covering difficult terrain without human troubleshooters on sight. How is a remote vehicle steered?

EXPLAIN A lunar rover or Moon rover is a space exploration vehicle (rover) designed to move across the surface of the Moon. Mars rovers have landed on the surface of Mars and are busily exploring the planet.

MOVE Place chairs in a mock travel bus formation and take a classroom virtual field trip. Project your tour on the wall or smart interactive board. As you "tour," students lean left, lean right, bounce in their seats, etc.

INVESTIGATE

Visit the Mars Rover site to interact with a virtual surface rover.
http://marsrover.nasa.gov/mission/spacecraft_surface_rover.html

Visit NASA Kids Club online.
http://www.nasa.gov/audience/forkids/kidsclub/flash/index.html

Take a Virtual Moon tour on Google Earth. Download here:
http://www.google.com/earth/explore/showcase/moon.html

ELA **Read the winning entries in Poetry Zone's "Space Poems" contest.**
http://poetryzone.woodshed.co.uk/spacecomp.htm

ROUND AND ROUND
STEMVESTIGATIONS: WHEELS

ENGINEERING

READ **Wheels on the Ground** *page 12*

DISCUSS What is a wheel? Are wheels all the same size? Are wheels found only in vehicles? Where else might wheels be located?

EXPLAIN Wheels are connected to an object on a rod called the axle like those on bikes, cars, or a set of rollerblades. Wheels are also used as doorknobs, clock gears, on pulleys, and in lots of mechanical appliances.

MOVE Drag a student seated in a box across the floor. Now pull students in the chair. Do a relay. One line of students will pull the box without wheels. One line of students pulls the box on wheels (a rolling chair). Which groups finishes first? Which group works harder? Attach a spring scale to show the amount of force applied to pull the box. (The scale will max out on the box pulled without wheels...)

BRAINSTORM

Student teams will use cut paper, recycled materials, string, thread, tape or other art supplies to make a vehicle that carries pencils across a tray of sand (simulate snowy ground or moon's surface). The goal is to make the least impression on the sand as possible. Drag, roll, pull, push, move 6 ounces of weight from one end of the sandy tray to the other.

- Fill a large brownie pan with sand (one inch deep at least).
- Supply recycled items and let students use classroom materials.
- Use clay for weight. Students must create the vehicle and container.
- Give a time limit, but let them get truly creative.
- Call time. Teams test their vehicles in the sand pit. Depth of mark left by the vehicle is criteria. Discuss why some vehicles sink or create more depression than others...

ELA Write and illustrate shape poems inside the circles.

HANDOUT PAGE 139

The ARTS

READ **Hanging with Circles** *page 14*

DISCUSS How did the author use the patchwork to talk about shapes and the alphabet? Could you make a pattern using only circles? What happens to the quilt edges?

EXPLAIN Circles placed side-by-side will not lay edge to edge. The curve of a circle leaves a gap between each shape. Even tiny circles placed next to each other will never entirely fill the gap. Circles can be overlapped, though to fill the background space.

MOVE Make people patterns. Give each student a round paper plate. Small teams of students (or the entire class) attempt to make many different patterns using the circles. Kids stand with the plates, using their arms to move the plates into patterns. Snap a quick photo with your phone or camera to review the unique patterns later.

INVESTIGATE

What unusual patterns can be created using circles? Trace circles of many different diameters, and use them to design unique mixed media creations. Combine this lesson with attributes. Students cut their circles from various textured papers and cloths (or glue feathers, glitter, etc. to the patchwork designs after "quilts" are created). Designs might be laminated to create place mats or artwork for parents.

ELA Create alphabet books featuring circles. Reimagine each animal as composed of circles! The zebra's hair has a circular pattern, circle for eyes, nostrils, etc..)

STEMVESTIGATIONS: RADIUS

MATHEMATICS

READ **Radius and Pi** *page 16*

DISCUSS "Yes, a line is fine, but when a line swerves, when a line bends, watch what can happen... a shape begins!" List and draw shapes already known.

EXPLAIN Circles are included in the oval category. All points on the bent line are the same distance from the center of the circle. That distance is called a radius. The entire distance across the circle is the diameter.

MOVE Take a shape walk. As a class, each student should have a blank sheet of paper on which is drawn a circle. Walk around the school and tally all the circles as they are discovered. Extend the lesson: create a map of the school and mark a dot on each circle found. Compare clusters. Where -in the school- are the most circles found? Hint: Keep the walk quiet. Students should make an "okay" sign with fingers to announce each circle's discovery.

INVESTIGATE

Measure the radius from the center of your head to the tip of your middle finger. (Measure from fingertip to fingertip to find the diameter of "your circle.") Cut a length of string measured to the length of "your radius." Pin "your radius" string to the middle of a large sheet of paper. With a pencil at the other end of the string, drag the string around in a complete circle.

ELA Write and illustrate shape poems inside the circles.

SQUARE, CIRCLE, TRIANGLE, RHOMBI!

reimagine the game of Duck, Duck, Goose

MATERIALS

Cut and laminate a variety of polygons.

SETUP

Seat students in a circle. Give each student a polygon. Choose a player to be "Polygon."

INSTRUCT THE PLAYERS

Students sit in a circle on the floor. Each student holds a shape. As "Polygon" walks around the circle, he or she touches each student's head and names the shape that student holds.

ex: circle, circle, triangle, pentagon, square, square

"Polygon" chooses a student and says, "RHOMBI."

Continue play as you would with a traditional game of "Duck, Duck, Goose."

UNIT 1

ROUND AND ROUND

Once upon a time, there was Rhombi. Rhombi loved shapes and found them everywhere. She especially loved the the circles she saw in the moon and sun.

On many nights, Rhombi sat in a meadow to watch the moon. She noticed that its shape appeared to change from week to week.

Rhombi wanted to reach out and touch the moon. She wanted to see if its shape really changed or if there might be some other explanation. Rhombi hopped on her bicycle to visit her Aunt. Rhombi's aunt was an astronomer. She studied objects in the universe.

Rhombi met her aunt on the porch. They looked through a big telescope that made the moon seem close enough to touch. It isn't, of course. The moon is 238,900 miles (384,400 km) from Earth. You could drive from one side of the USA to the other more than 79 times to equal that distance. "The moon," said Rhombi's Aunt, "moves away from Earth about one and a half inches every year."

Rhombi asked, "Why does the moon change shape?"

Aunt laughed, "It stays the same shape. We see an Earth shadow that makes it look different from week to week. Why don't you borrow my telescope for the month? I'm headed to Mexico to use The Very Large Array in New Mexico."

ROUND AND ROUND

Rhombi was excited. Her Aunt told Rhombi that she could get the telescope anytime. Rhombi wanted to draw the moon every day to see how the Earth's shadow affected the moon. She did not have any way to pick up the telescope, though. She looked out from the pages of her world and asked for help. "Designers, will you help me create a wagon to move the telescope to my house?"

RHOMBI'S ADVENTURES IN 2D
TALK ABOUT S.T.E.A.M.

Can you reach out and touch the moon?

The Earth has a Diameter of 7,918 miles. So when you make a real distance measurement. Remember that the Moon is 30.1 Earths away. Lay a basketball on the ground. Mark off 30 basketballs and place a tennis ball at the end of the line. This demonstration is a fairly accurate measure of the moon's distance from Earth.

Watch a great video at http://www.youtube.com/watch?v=Bz9D6xba90g

Do you think the moon changes shape each week?

The moon does not actually change shape. Earth casts a shadow on the moon that makes the planetary object appear to change..

What tool will Rhombi use to watch the moon?

A telescope is the right tool.

What does Rhombi's aunt do for a living?

She is an astronomer.

Where is Rhombi's aunt traveling?

She's going to Mexico.

What will Aunt use in Mexico?

She will use The Large Millimeter Telescope on Sierra Negra in Mexico.

What does the moon look like from week to week?

crescent, half, full, half, crescent

Make a grid to chart the moon's appearance for one month.

RHOMBI'S ADVENTURES IN 2D
UNIT 1 CHALLENGE: ROUND AND ROUND

"Designers, will you create a 3 inch by 3inch square cart to move my things across the ground?"

1. The cart will need wheels that rotate around an axle.
2. The cart you build will need to carry 10 ounces of material for a distance of 12 feet without falling apart.

My full size cart will need to carry 1000 ounces over several miles. You are making a 1:100 scale model design of my cart. That means all measurements on your scale model classroom cart will measure 100 times greater when we build the real life-size cart.

Draw a wheel and axle.

Draw your cart design.

Draw each of the pieces you will need to complete the challenge.

wheel?

axle rod?

way to keep wheels on the rod?

cart shape?

On the back of this paper, write out the steps you will need

RHOMBI'S ADVENTURES IN 2D
RUBRIC: ROUND AND ROUND

TASK **Students will use their knowledge of circles to troubleshoot and build a non-motorized cart that rolls.**

	Content	Organization	Presentation
1	• Is well thought out and supports the solution to the challenge or question • Reflects application of critical thinking • Has clear goal that is related to the topic • Is accurate	• Information is clearly focused in an organized and thoughtful manner • Information is constructed in a logical pattern to support the solution	• Format enhances the content
2	• Supports the solution • Has application of critical thinking that is apparent • Has no clear goal • Has some factual errors or inconsistencies (i.e. wheels do not rotate around axle)	• Project has a focus but might stray from it at times (more concerned with form than function) • Information appears to have a pattern, but the pattern is not consistently carried out in the project	• Format does not suit the content
3	• Provides inconsistent information for solution • Has no apparent application of critical thinking • Has no clear goal • Has significant factual errors, misconceptions, or misinterpretations	• Content is unfocused and haphazard • Information does not support the solution to the challenge or question • Information has no apparent pattern	• Presentation appears sloppy and/or unfinished

RHOMBI'S POLYGON ROUND-UP
CIRCLE

ABOUT THE CIRCLE

- In the math world, circle refers to the boundary of the shape. "Disk" is used to refer to the whole shape, including the inside.
- A straight line from the center of a circle to the edge is called the radius.
- A straight line that passes from one side of a circle to the other through the center is called the diameter.
- The distance around the outside of a circle is called the circumference.All points on the edge of a circle are the same distance to the center.
- The value of Pi (π) to 2 decimal places is 3.14, it comes in handy when working out the circumference and area of a circle.
- Circles have a high level of symmetry.

UNIT 2
LOOKING FOR SQUARES

LOOKING FOR SQUARES

Essential Questions

1. What are the attributes of a regular square?
2. How does technology help you learn about new things?
3. What strategies make counting easier?
4. When would a pattern make a task simpler?
5. What are you trying to accomplish in Rhombi's Challenge?

ENDURING UNDERSTANDING

Two dimensional shapes create new shapes.

"Halves" means two equal parts."

Circles and rectangles break into equal parts.

Breaking circles or rectangles into more equal parts means that the parts will be smaller.

Three objects may be put in order from longest to shortest by comparing their lengths.

The length of an object may be discussed using whole numbers.

A smaller object may be used as a measurement tool when measuring longer lengths.

Larger numbers consist of tens and ones.

Organizing information helps in discussion.

Focus on quadrilaterals, trapezoids, parallelograms, squares and rectangles. Use tallies to "harvest" shapes at lunch. Employ technology in discovering and discussing geometry. Sharpen observational skills with shape walking. Research and recreate famous quilted art. Make your own origami square papers, and learn to fold boxes.

Students will apply basic measurement, understanding of number, place value, data collection, and data application in problem solving.

Rhombi needs to know which shapes in her school are actually squares, and which just look squar-ish. Students design a tool to check the "squareness" of a shape.

Watch for these extension level words:

builder
In this case, the term "builder" refers to a person that creates or develops a particular thing.

contractor
a manager and coordinator of all the tradespeople involved in a project of many phases

designer
a person who plans the form, look, or workings of something before its being made

engineer
a person who has scientific training and who designs and builds complicated products, machines, systems, or structures : a person who specializes in a branch of engineering

LOOKING FOR SQUARES
CURRICULUM CONNECTIONS

Rhombi needs a tool to determine square shapes.

MATH BACKGROUND - Circles

Squares are -by definition- regular polygons in that all edges must be the same length. Because they have 4 sides of equal length, squares are regular quadrilaterals. Squares are also rectangles with equal sides and rhombuses with right angles. Square shapes are often used by for design and engineering purposes such as town planning.

TECHNOLOGY CONNECTION

Build logic and spatial reasoning.

- www.mathplayground.com/tetrasquares.html
- www.mathsisfun.com/games/magic-square-game.html

Mr. R's Math Videos and Songs

- www.youtube.com/user/mathsongs1/videos

HISTORICAL CONNECTIONS

The process of quilting uses a needle and thread to join two or more layers of material. During pioneer days, "paper" quilting became popular. Each piece of cut fabric was basted around paper patterns. Paper was scarce in the west, and women would save letters from home, newspaper clippings, and catalogs to use as patterns. The paper acted as pattern and insulator. Paper found between old quilts is a primary source of information about pioneer life. During the Civil War, slaves used quilts as a means to share and transmit secret messages.

ART CONNECTIONS

Piet Mondrian was born in Amersfoort in the Netherlands, in 1872. Mondrian began his career with traditional art styles, but embrace the concept of "cubism." His most famous painting was "Composition with Red, Yellow and Blue" painted in Paris in 1921.

Cubism was developed by Pablo Picasso and George Braque in the beginning of the 20th century. A typical Cubist painting depicts real people, places or objects, but not from a fixed viewpoint. Instead it will show you many parts of the subject at one time, viewed from different angles, and reconstructed into a composition of planes, forms and colours.

READING CONNECTION

Introduce each activity and lesson with a book. Books are listed on the Read about STEAM page. Jot additional lists of books from your favorites here.

1.

2.

3.

4.

LOOKING FOR SQUARES
CURRICULUM CONNECTIONS

COMMON CORE CONNECTIONS

ELA/Literacy
W.K.7 Participate in shared research and writing projects (e.g., explore books by a favorite author and express opinions). (K-ESS2-1) add venn

Mathematics
MP.2 Reason abstractly and quantitatively. (K-ESS2-1)

MP.4 Model with mathematics. (K-ESS2-1)

K.CC.A Know number names and the count sequence. (K-ESS2-1)

K.MD.A.1 Describe measurable attributes of objects, such as length or weight. Describe several measurable attributes of a single object. (K-ESS2-1)

K.MD.B.3 Classify objects into given categories; count the number of objects in each category and sort the categories by count. (K-ESS2-1)

NEXT GEN SCI CONNECTIONS

ETS1.A: Defining Engineering Problems
A situation that people want to change or create can be approached as a problem to be solved through engineering. Such problems may have many acceptable solutions. (secondary to K-PS2-

1-ESS1-1,1-ESS1-2 Patterns in the natural world can be observed, used to describe phenomena, and used as evidence.

1-ESS1-1, 1-ESS1-2 Make observations (firsthand or from media) to collect data that can be used to make comparisons. Analyzing data in K–2 builds on prior experiences and progresses to collecting, recording, and sharing observations. Use observations (firsthand or from media) to describe patterns in the natural world in order to answer scientific questions.

RESOURCES: UNIT 2: S.T.E.A.M. ACTIVITIES "LOOKING FOR SQUARES"

Rhombi Audio Download / Video (available April 2014)

UNIT 2:	Looking for Squares:	Science:	Look for shapes with food.
UNIT 2:	Looking for Squares:	Technology:	Use digital photography to interact with shapes.
UNIT 2:	Looking for Squares:	Engineering:	How can you determine that a shape is a square?
UNIT 2:	Looking for Squares:	Art & Music:	Discover the geometry of quilts.
UNIT 2:	Looking for Squares:	Mathematics:	Explore paper squares.
UNIT 2:	Looking for Squares:	ELA:	Read "Looking for Squares," and illustrate.

LOOKING FOR SQUARES
BUILD S.T.E.A.M. WITH GREAT BOOKS

SCIENCE

Pick a Circle, Gather Squares: A Fall Harvest of Shapes
by Felicia Sanzari

SQUARES (Shapes in Math and Nature)
by Sheldrick Ross, Catherine and Bill Slaven

TECHNOLOGY

The Kids' Guide to Digital Photography: How to Shoot, Save, Play with & Print Your Digital Photos
by Jenni Bidner

Looking Through a Telescope
by Linda Bullock

ENGINEERING

Finding Shapes with Sebastian Pig and Friends at the Museum
by Jill Anderson

Shapes : squares
by Sarfatti, Esther

Shapes at School
by Joyce Jeffries

ART & MUSIC

Art & Lots of Dots
by Ginett Alarcon

Dare to Be Square Quilting: A Block-by-Block Guide to Making Patchwork and Quilts
by Boo Davis

MATH

Sam's Sneaker Squares
by Math Matters & Nat Gabriel

My Mother Taught Me Origami
by Manorama Jafa

ELA

Square Bear Meets Round Hound
by Marianne W. Penn

Doodle Dandies: Poems That Take Shape
by J. Patrick Lewis

SCIENCE

READ **Square Foot Harvest** *page 20*

DISCUSS What do students already think they know about harvesting? Make a list to identify "mindbugs."

EXPLAIN When a crop has grown and is ready to eat, the farmer must collect it from the fields. This is known as the harvest. Grain crops are collected by huge combine harvesters. Some tree fruits are picked by tall machines. Various crops are harvested all year.

MOVE Pretend to harvest various foods. Will you stretch tall to reach an apple? Should you bend deep to pull carrots or peanuts? Call out a food, and students move through the motions.

INVESTIGATE

"Harvest" the shapes you can find at lunch. What vegetables and fruits are present in student lunch boxes or on their cafeteria trays? Work with students to create a chart for tallying the various foods. Students should offer their educated guesses as to which foods will be represented. After tallying lunch items, compare the guesses with the actual tallies. This is an opportunity to combine counting, data representation, data analysis, geometry and data collection in a simple activity.

ELA Imagine the harvest using five senses. Describe colors, shapes, tastes, sounds, and textures. Students close their eyes and list a descriptive word or phrase per sense.

STEMVESTIGATIONS: DIGITAL FIELD TRIP

TECHNOLOGY

READ **P Is for Parallel** *page 22*

Curious George Dinosaur Tracks by H. A. Rey

DISCUSS Digital photography includes all those pictures taken on digital cameras, tablets and mobile phones. How many students take digital photos? Why is George snapping pictures? What is the story's mix-up?

EXPLAIN Digital photography uses light sensitive electronic devices known as an image sensors. These sensors are made up of millions of photoelectric devices that convert light into an electrical signal.

MOVE Take a mobile phone digital photography walk. Let kids point out objects worthy of a photo. Each time they name an object, the student must also explain what it is using the vocabulary of geometry.

INVESTIGATE

Once back in the classroom, give a slideshow of photos taken and create a PowerPoint presentation of square and quadrilateral shapes found in the school.

ELA "Blog" about the journey around your school and what was found on the investigative hike.

LOOKING FOR SQUARES
STEMVESTIGATIONS: SQUARE HUNT

ENGINEERING

READ **Shopping for Shapes** *page 24*

DISCUSS From chairs to skyscrapers, every object has a shape. Often, several shapes are hidden in plain sight. Brainstorm a list of locations at which kids see shapes every day.

EXPLAIN Squares have uses in architecture, engineering, and technology. The square must have edges of equal length and 4 right angle corners (vertices).

MOVE Use bodies, arms, legs and books to make shapes as they are called out by "Rhombi". Rhombi says, "Show me a square." Rhombi says, "Make your arms into a circle."

INVESTIGATE

What squares can be found in your school? Create a comprehensive list of square objects and objects that incorporate squares. How will you check for square shape vs. not? Not all things that appear identical are really the same. Remember to look up at the ceiling and down at the floor when square hunting.

ELA Make a diorama of your school or a classroom using cut shapes and gluesticks.

STEMVESTIGATIONS: QUILT PATTERNS

The ARTS

READ **Stories on the Square** *page 26*

DISCUSS Quilts are generally geometric designs created using fabric shapes. Quilt patterns may be simple repeating squares or circles. Patterns may also be complicated.

EXPLAIN Quilts use all shapes as part of images sewn to add interest. Modern quilts have much brighter fabrics than older quilts and often boast more boldly geometric patterns.

MOVE Bring a quilt (or blanket) to play. Students stand around the edge of the square quilt holding its edges with both hands. Pick up the quilt. The teacher drops a balloon on the quilt, and students must work together to pop their balloon up into the air and keep it on the quilt.

INVESTIGATE

Using patterns from a quilting source, use paper to cut-and-paste quilts instead of sewing them. Recreate famous quilt patterns, research modern patterns, and use the resulting geometric artwork as a way to discuss patterns, shapes, measurement, and organization.

ELA Write fictional stories about the paper quilt owners.

STEMVESTIGATIONS: PAPER SQUARES

MATHEMATICS

READ **Paper Forests** *page 28*

DISCUSS What is origami? Origami is the art of folding paper to look like specific shapes, animals, boxes, and other items.

EXPLAIN Origami papers are shaped like squares, because that shape offers the widest variety of symmetrical folds. Both halves of any origami shape are identical. If the origami bird has a left wing, it will also have a right wing.

MOVE Fold a square! Fold a triangle! Begin with any piece of notebook or construction paper. Fold the top corner down so that A and AA line up exactly. Cut along the resulting line to remove excess paper. Unfold the triangle to reveal a regular square.

INVESTIGATE

Make your own origami square papers. What shapes can you fold from the starting point of a simple square? Experiment with various paper thicknesses. What happens when you fold the same shape from three different types of paper (construction, lined, tissue)? Students should journal about each paper investigated, its attributes, and their observations about its use in origami.

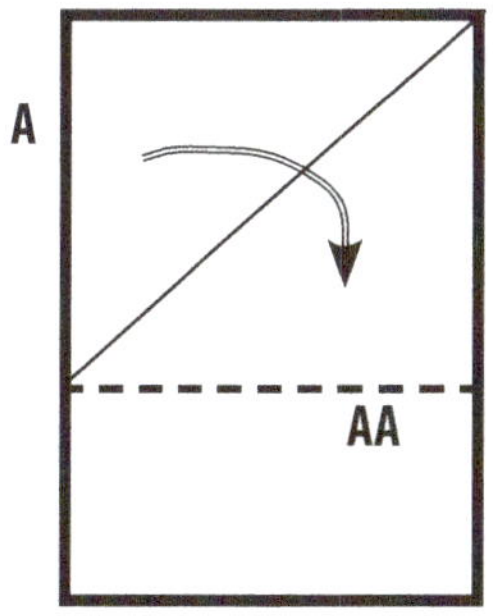

ELA Compile student journal entries into a group guide.

HOW MANY EDGES DOES IT HAVE?

MATERIALS

Objects with faces including circle, triangle, square, rectangle, pentagon.
Blindfold

SETUP

Gather shapes OR cut and laminate shapes for each group.

INSTRUCT THE PLAYERS

Students work in pairs or small groups. One student is designated "Rhombi." Rhombi will wear a blindfold and try to name the shape using only touch. Before naming the shape, Rhombi must name the number of edges and the number of vertices. Player earns a point for the complete correct answer. Rotate in a clockwise direction to the next player.

STUDENTS BREAK INTO TWO CLASS TEAMS OR SMALL TEAMS

Each team gets a pile of objects and a blindfold (baggies will keep everything neat). While students play, monitor the discussions. Players should be naming the number of edges and vertices.

UNIT 2

LOOKING FOR SQUARES

Once upon a time, there was Rhombi. Rhombi loved to look for shapes. Rhombi's aunt had given her a telescope. Rhombi had used her innovative new wagon to carry the tool around town.

Rhombi loved the circular shape of the moon when it was full. She wondered what other shapes were hiding in plain sight. The geo-explorer grabbed her notebook, and headed into the hallways.

Rhombi decided to take a walk around the school to find more shapes. She looked up at the big round sun for just a second. She knew not to stare at the sun. Inside her school, Rhombi found a round clock on the wall.

Rhombi found square-looking shapes in her classroom. Floor tiles looked like squares. Ceiling tiles looked like squares. Even the math book looked like a square. Rhombi knew that not all square-looking things were really squares.

LOOKING FOR SQUARES

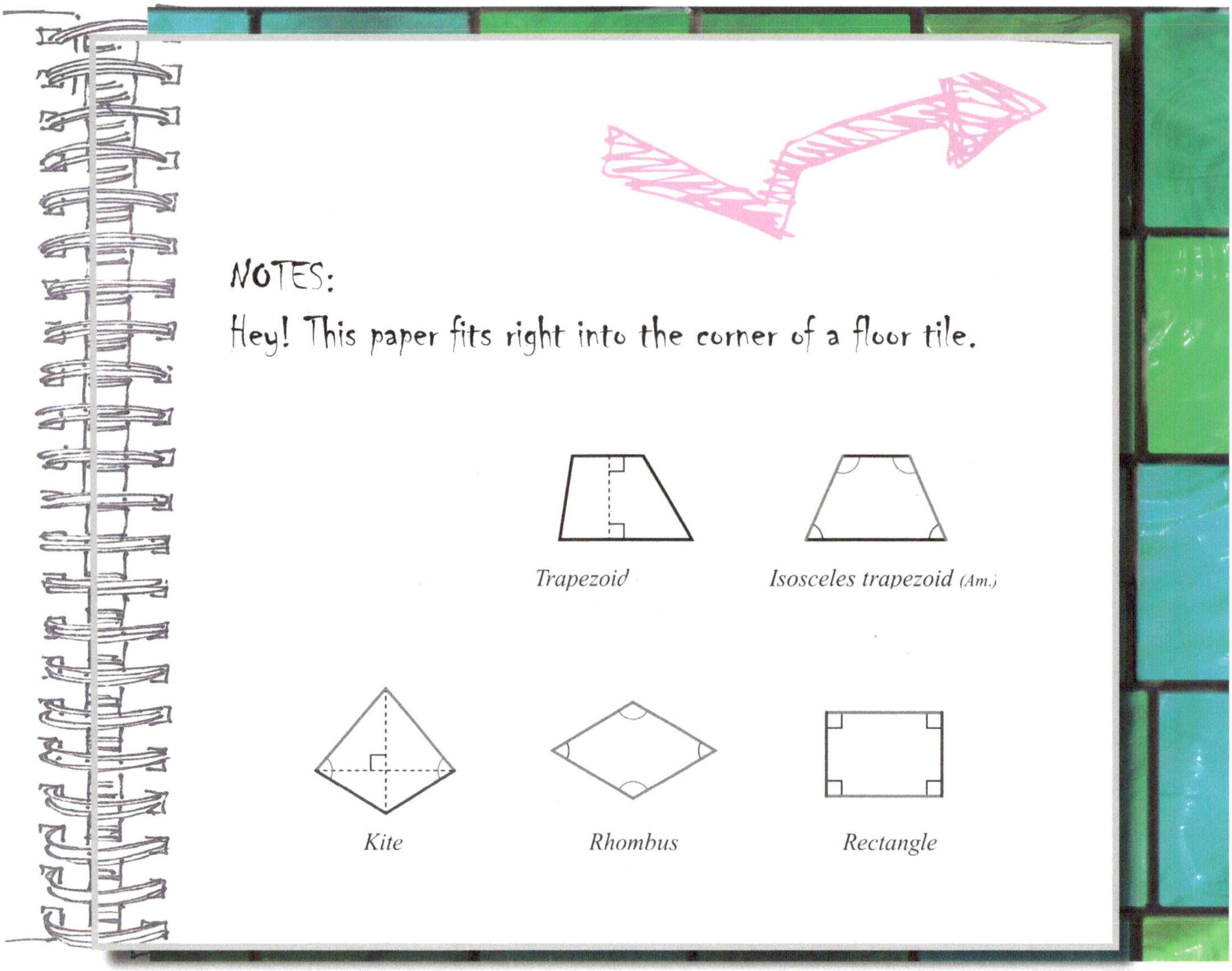

Rhombi took a closer look at the tiles. Rhombi knew that a square had four sides of the same length. They had four vertices at right angles. Rhombi pushed a page of her notebook into the corner of the tile. It fit smoothly. Interesting!

LOOKING FOR SQUARES

Rhombi needed a tool to decide which things were squares and which were just quadrilaterals and trapezoids. Rhombi looked out from the pages of her world and asked, "Designers, can you create a tool to help identify squares?"

LOOKING FOR SQUARES
TALK ABOUT S.T.E.A.M.

What gift did Rhombi give her? **telescope**

What vehicle did Rhombi use to transport her telescope? **wagon**

What do you think a geo-explorer is...? **Someone that looks for shapes...**

List some round objects found by Rhombi. **clock, sun**

Why shouldn't you stare at the sun? **The light from the Sun can severely damage your eyes.**

What classroom objects looked like squares? **ceiling and floor tiles**

What is Rhombi trying to learn? **How do you verify that an object is square?**

What tool does Rhombi need? **identify squares**

Can you name some things that have corners like a square?
paper, ceiling and floor tiles

What is a vertice? What does a right angle look like?
A vertice is a corner. A right angle looks like an L.

How do you know if a shape is a quadrilateral? **4 sides / edges**

How do you know if a shape is a trapezoid? **parallel set of sides**

How could you tell a rectangle from a square? **A square is a rectangle, but a rectangle is not a square. Squares have 4 sides of equal length and 4 90 degree angles.**

RHOMBI'S ADVENTURES IN 2D
UNIT 2 CHALLENGE: LOOKING FOR SQUARES

"Designers, will you create a tool to help me identify squares?"

1. The tool must show a square's L shaped corner angles.
2. The tool must test squares of all sizes.
3. The tool must be mobile (can be carried around).

Hint: *Regular paper has a 90 degree angle shaped like an L.*
Hint: *Origami can help you cut a square for your pattern and help you find the center of a square.*

Draw a square. Measure to make sure that all 4 sides are the same length. Make sure the corner angles are shaped like an L at 90 degrees.

Draw your tool design.

Draw each of the pieces you will need to complete the challenge.

paper?
scissors?
string or
ruler?

On the back of this paper, write out the steps you will take to complete your task.

RHOMBI'S ADVENTURES IN 2D
RUBRIC FOR LOOKING FOR SQUARES

TASK **Students will use their knowledge of squares to design and execute a tool for identifying squares.**

	Content	Organization	Construction
1	• Is well thought out and supports the solution to the challenge or question • Reflects application of critical thinking • Has clear goal that is related to the topic • Is accurate	• Information is clearly focused in an organized and thoughtful manner • Information is constructed in a logical pattern to support the solution	• Outcome has clean construction and fits the challenge performance requirements. • Object is accurate.
2	• Supports the solution • Has application of critical thinking that is apparent • Has no clear goal • Has some factual errors or inconsistencies (i.e. wheels do not rotate around axle)	• Project has a focus but might stray from it at times (more concerned with form than function) • Information appears to have a pattern, but the pattern is not consistently carried out in the project	• Outcome has been carefully constructed to the best of the student's ability but lacks solidity. • Object is moderately accurate.
3	• Provides inconsistent information for solution • Has no apparent application of critical thinking • Has no clear goal • Has significant factual errors, misconceptions, or misinterpretations	• Content is unfocused and haphazard • Information does not support the solution to the challenge or question • Information has no apparent pattern	• Outcome is sloppy and hastily constructed. • Outcome is inaccurate.

RECTANGLE

ABOUT THE SQUARE

- A square is a polygon with 4 sides of equal length and 4 right angle corners (90 degree corners).
- Because it has 4 sides of equal length, a square is a regular quadrilateral.
- A square is also a rectangle with equal sides and a rhombus with right angles.
- The perimeter of a square is 4 times the length of one side.
- Opposite sides of a square are parallel.
- The internal angles of a square add to 360 degrees.
- A square has 4 lines of reflectional symmetry.

UNIT 3

TASTY TRIANGLE

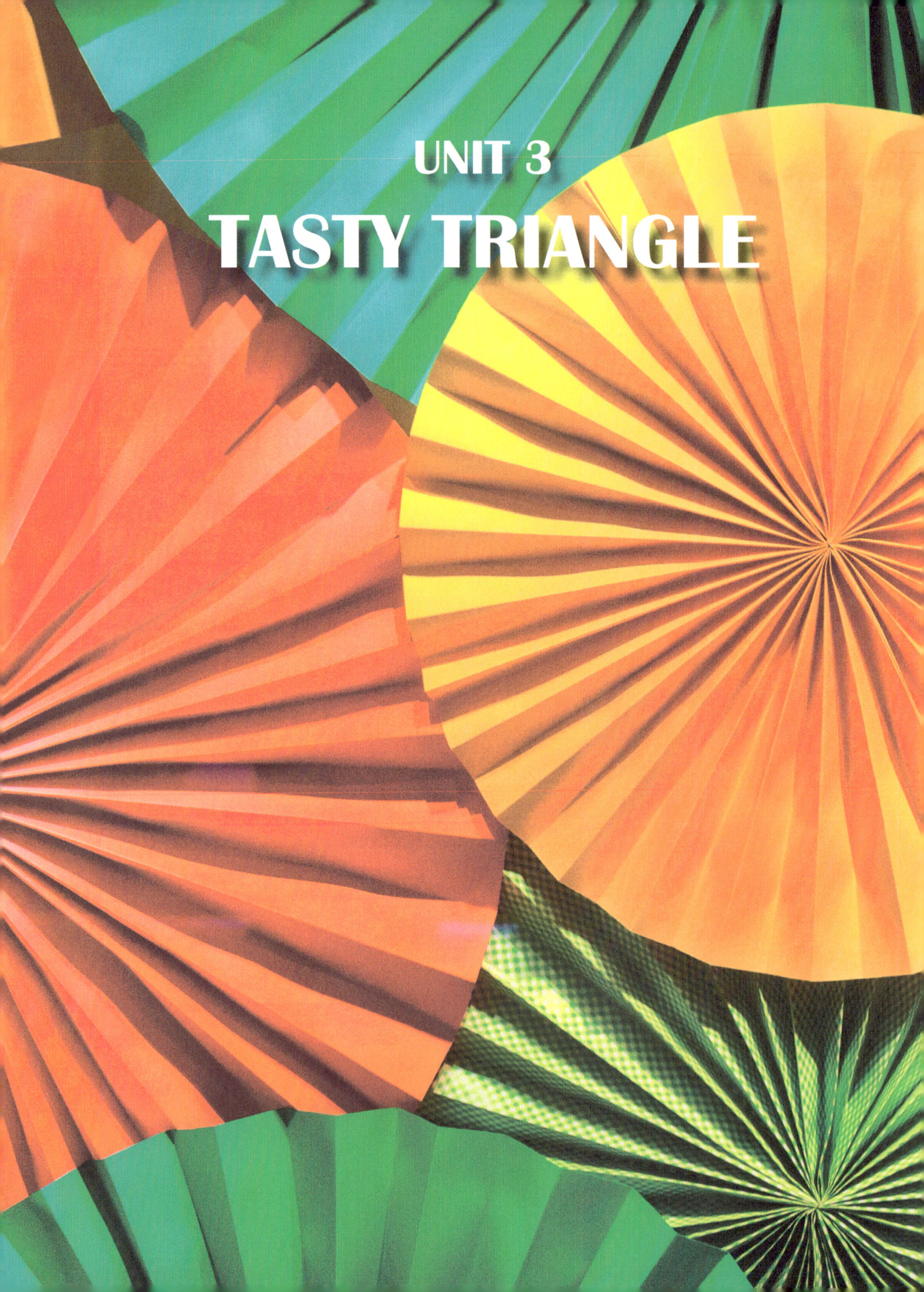

TASTY TRIANGLE

Essential Questions

1. How many edges and vertices are present in a regular triangle?
2. Where are triangles found in the building?
3. Why do engineers use triangles in building bridges and other structures?
4. Where would you find triangles in the world?
5. What are you trying to accomplish in Rhombi's Challenge?

ENDURING UNDERSTANDING

Two dimensional shapes create new shapes.

The length of an object may be discussed using whole numbers.

Organizing information helps in discussion.

Number information may be organized for the purpose of answering questions.

Explaining number information makes more sense when it is organized.

Shapes may be identified by their attributes.

Focus on triangles, angles, regular triangles, and corners (vertices). Find apps and games to create triangles on the Internet. Improve Rhombi's cart. Explore primary and secondary colors. Create a triangular world as digital images are remade with triangle overlays.

Students will develop extended measurement skills while deepening understanding of two dimensional attributes. Connect with historical backgrounds and introduce tesselations if possible. Watch Vi Hart do some Triangle Doodling. Her videos are as entertaining as they are educational.

Help Rhombi design a folded paper bridge for the mice.

origami
Japanese art of folding paper

TASTY TRIANGLE
CURRICULUM CONNECTIONS

Rhombi needs a bridge supported by triangles.

MATH BACKGROUND - Triangles

A triangle is one of the basic shapes in geometry: a polygon with three corners or vertices and three sides or edges which are line segments.

TECHNOLOGY CONNECTIONS

Check out the Math Munch by Vi Hart.
http://mathmunch.org/tag/vi-hart/

Khan Academy's Triangle Doodling by Vi Hart.
http://alturl.com/ntx8p

HISTORICAL CONNECTIONS

Triangles are present in architecture dating to earliest recorded history. Examples include:

- Egypt- Great Pyramids of Giza
- Athens- Parthenon
- Spain- Palais Ideal
- USA - Golden Gate Bridge

Unlike a rectangle, a triangle cannot be deformed without changing the length of one of its sides or breaking one of its joints. In fact, one of the simplest ways to strengthen a rectangle is to add supports that form triangles .

ART CONNECTIONS

9 Pointed Star
This work of art is made up of 16 triangles. Each triangles is equilateral and they rotate to the same degree each time. The triangles are stacked on top of each other. The triangles are used to make a unique star with nine points.

Tilings or tesselations refers to the complete covering of a plane surface by tiles. Pompeii Tiling with equilateral triangles was an example of tesselation.

Go online to create tesselations at this site:
http://gwydir.demon.co.uk/jo/tess/tri.htm

READING CONNECTIONS

Introduce each activity and lesson with a book. Books are listed on the Read about STEAM page. Jot additional lists of books from your favorites here.

1.

2.

3.

4.

TASTY TRIANGLE
CURRICULUM CONNECTIONS

COMMON CORE CONNECTIONS

ELA/Literacy
W.K.7 Participate in shared research and writing projects (e.g., explore books by a favorite author and express opinions). (K-ESS2-1) add venn

Mathematics
MP.2 Reason abstractly and quantitatively. (K-ESS2-1)
MP.4 Model with mathematics. (K-ESS2-1)
K.CC.A Know number names and the count sequence. (K-ESS2-1)
K.MD.A.1 Describe measurable attributes of objects, such as length or weight. Describe several measurable attributes of a single object. (K-ESS2-1)
K.MD.B.3 Classify objects into given categories; count the number of objects in each category and sort the categories by count. (K-ESS2-1)

NEXT GEN SCI CONNECTIONS

K-2-ETS1-1. Ask questions, make observations, and gather information about a situation people want to change to define a simple problem that can be solved through the development of a new or improved object or tool.

K-2-ETS1-2. Develop a simple sketch, drawing, or physical model to illustrate how the shape of an object helps it function as needed to solve a given problem.

K-2-ETS1-3. Analyze data from tests of two objects designed to solve the same problem to compare the strengths and weaknesses of how each performs.

K-PS2-1. Plan and conduct an investigation to compare the effects of different strengths or different directions of pushes and pulls on the motion of an object.

K-PS2-2. Analyze data to determine if a design solution works as intended to change the speed or direction of an object with a push or a pull.*

RESOURCES FOR UNIT 3: S.T.E.A.M. ACTIVITIES FOR "TASTY TRIANGLE

Rhombi Story Book
Rhombi Audio Download / Video (Available December 2014)

UNIT 3: TastyTriangle:Science:	**Find the strongest materials for triangle constrction.**
UNIT 3: TastyTriangle: Technology:	**Use the apps and Internet sites to interact with shapes.**
UNIT 3: TastyTriangle: Engineering:	**Design and improve Rhombi's mode of travel.**
UNIT 3: TastyTriangle: Art & Music:	**Use triangles to draw just about anything.**
UNIT 3: TastyTriangle: Mathematics:	**Reshape the world with triangles.**
UNIT 3: TastyTriangle: ELA	**Read "TASTY TRIANGLE." Students illustrate the story.**

TASTY TRIANGLE
BUILD S.T.E.A.M. WITH GREAT BOOKS

SCIENCE

Showdown at the Food Pyramid
by Rex Barron

Triangles (Shapes in Math, Science and Nature)
by Catherine Sheldrick Ross and Bill Slavin

TECHNOLOGY

The Wing on a Flea
by Ed Emberley
http://www.youtube.com/watch?v=0-YmXsqq_WU

Triangles and Pyramids
by Sally Morgan

"The Polygon Song"
by Steve Saunders
http://youtu.be/TTrbT_SSD0U

ENGINEERING

Triangles Around Town
by Nathan Olson

The Bridge
by Emily Cheney Neville

The Art of Construction
by Mario Salvadori

ART & MUSIC

Triangle for Adaora: An African Book of Shapes
by Ifeoma Onyefulu

Draw a Triangle, Draw Anything!
by Chris Hart

MATH

Triangles
by Bill Slavin, Catherine Sheldrick Ross

Mouse Shapes
by Ellen Stoll Walsh

ELA

The Greedy Triangle
by Marilyn Burns

Pop's Bridge
by Eve Bunting

STEMVESTIGATIONS: GROWING POLYGONS

SCIENCE

READ **Rickety Rackety** *page 32*

DISCUSS Where do you see triangles?

EXPLAIN Polygons may have three, four, five, six, or more sides. All sides must be straight. Triangular polygons are more stable than other polygon shapes. The angles in a triangle cannot change once the triangle is built. The angles are fixed. Because a triangle has three sides and three angles, each angle is fixed by the side opposite to it.

MOVE Use chalk and straight edges to mark regular polygons on the sidewalk. Use chalk and straight edges to mark irregular polygons on the sidewalk.
regular triangle, regular square, regular pentagon, regular hexagon
Irregular triangle, irregular square, irregular pentagon, irregular hexagon

INVESTIGATE
Use straws and clay (or marshmallows) to construct regular polygons in the shapes of regular triangles, regular squares and regular pentagons. Flex each polygon. Which one seems least able to bend or change its corners (angles)? Compare the strength and stability of each shape. Make polygons from many different materials. Use clay, marshmallows, foam, gum and any other connector to prove that the shape leads to the triangle's strength rather than any fixative. Keep careful records, and make sure to use the same amount of material in each test run.

ELA Create a journal entry regarding the "tests" performed on each material. Present findings to the class.

TASTY TRIANGLE

STEMVESTIGATIONS: PRESENTATIONS

TECHNOLOGY

READ **City of Three Angles** *page 34*

DISCUSS Angles are the measurements used to define shapes in geometry. Triangles are described by their angles and number of equal edges.

EXPLAIN Triangles are grouped in three categories: acute angles are smaller than an "L" shaped angle. Right angles are "L" shaped angles. Obtuse angles are more open than "L" shaped angles.

MOVE Use arms to form acute, right and obtuse angles. Make up a class dance for triangles!

INVESTIGATE

Visit NCTM to play with triangles on the interactive GeoBoard.
http://www.nctm.org/standards/content.aspx?id=25007

Visit MathCats to interact with Polygons.
http://www.mathcats.com/explore/polygons.html

Make symmetrical pictures at Kali.
http://www.scienceu.com/geometry/handson/kali/

ENGINEERING

READ **Traveling Triangle** *page 36*

DISCUSS Traveling Triangle describes several places to find triangles. What do all the structures have in common? They all make structures stronger and more stable. What other triangles can you find in walls, fences, chairs, tables, etc?

EXPLAIN Triangular polygons are more stable than other polygon shapes. The angles in a triangle can not change once the triangle is built. The angles are fixed. Because a triangle has three sides and three angles, each angle is fixed by the side opposite to it.

MOVE Look for examples of triangles making structures strong. Students play a version of "musical chairs" using triangles in objects. Play music while students look for examples. Stop music. Anyone left without an example must sit down...

INVESTIGATE

Challenge students to build a platform using 4 blocks and cardboard. How much weight may be balanced on the platform before it topples?

Challenge students to build a cardboard platform held up by 3 blocks in a triangular formation. How much weight may be balanced on the platform before it topples?

ELA Create a journal entry regarding the "tests" performed on each material. Present findings to the class.

HANDOUT PAGE 149

The ARTS

READ **Fox's Forest** *page 38*

DISCUSS How many objects can be drawn using triangles? Where do you see triangles in faces? Birds? Lizards?

EXPLAIN The triangle does not exist as a regular polygon on animals. Triangles are something we've created, but we can recognize the essential shape using our imaginations.

MOVE Laminate a triangle (and other shapes). One team of students should afix the shape to an object in the classroom that contains a triangle. Other teams attempt to locate the shape first. Play progresses as teams take turns placing the triangle (and other shapes) until reaching 180 points. 180 is the total sum of all three angles inside a triangle.

INVESTIGATE

There are three primary paint colors (blue, red, yellow), the three pigments that form all art. Photocopy a regular equilateral triangle for each student. Paint one edge blue. Paint one edge red. Paint one edge yellow. Investigate the blending of primary colors. Using a paintbrush, gently pull some of each color toward the triangle's center. What colors result from blending all three? These colors are called secondary colors.

ELA Make lists of objects in each color group. Describe emotions that might correspond to the various colors.

MATHEMATICS

READAbout

READ **Tri, Tri Again** *page 40*

DISCUSS Watch the animation based on Ed Emberley's "The Wing on a Flea," and discuss the shapes seen in nature. Remind students that these shapes are not exactly like the regular shapes students will see on tests.

Animation via http://youtu.be/0-YmXsqq_WU
Learn about Emberley via http://youtu.be/zqcsYOSL_vo

EXPLAIN Geometric shapes used in mathematics have very specific rules to keep all mathematicians on the same page. Triangles must have corners (angles) with measurements that add up to exactly 180 degrees. Regular triangles must have edges / sides that are all the same length.

MOVE Take a digital shape walk. Snap photos of various scenes around the school that look interesting shape-wise.

INVESTIGATE

Using your digital images of the school, choose a few for students to use. In a presentation software program, have students place shapes over parts of the picture. For example, place a triangle over the angled rooftop. As each group takes a turn with the image, they will add their own shape replacements. Eventually, your class will create a fully re-shaped image!

Free versions to download:

Empressr
http://www.empressr.com/

Prezi
http://prezi.com/

SHAPE MIMES

MATERIALS

Pictures of shapes or shape names written on paper

SETUP

Break students into teams of 3 or 4 students. Place shape pictures or names in a bag or hat. Create a points tally chart on the board.

INSTRUCT THE PLAYERS

Teams will compete against one another for points OR adds points team-by-team to the whole class goal. Choose team names. One person from the first team chooses a shape picture / name from the bag. The team members must convey to students which shape they've picked BUT MAY NOT SPEAK AT ALL.

- Point to an example in the room.
- Make the shape with your arms.
- Form the shape with your bodies (3 kids at the vertices of a triangle, etc.).
- Trace the shape in the air.

UNIT 3

TASTY TRIANGLE

Once upon a time there was Rhombi. Rhombi loved shapes. Circles were cool (there were circles in the word "cool"). She had also had help figuring out how to test for squares, and that was a lot of fun.

Rhombi's penpal from Japan had sent her some square origami paper and instructions for making a folded fan. First, she accordion folded the longest side of the paper. Rhombi was interrupted by a tiny commotion near the stream.

3 tiny mice were trying to haul a box of raisins across the park to their mouse hole. They were stuck with no way to cross the little stream. Rhombi wished she could help, but they wouldn't let her pick them up.

When Rhombi returned to her park bench, someone had dropped the Wednesday newspaper on top of her fan. Oh no! Rhombi expected the fan to be smashed flat, but it was not smashed at all.

The folded paper was holding up the newspaper all by itself. Rhombi bent down to look more closely at the paper. She saw triangles. Once folded, the paper fan had made strong triangles with the bench as each triangle's base.

Rhombi looked around the park. She saw triangles in the bridge. She saw triangles in the park swings. Rhombi even saw triangles holding up the bicycles ridden in the park.

Rhombi thought about the little mice and their impossible task. She wished the little guys had a bridge. Wait! Rhombi new how to help the mice. She looked down at all the origami paper in her hands. The mice needed a bridge.

Rhombi would need help, and she knew just where to find the helpers. She looked out beyond the pages of her world and asked, "Designers, can you create a folded paper bridge that will help the mice get their box home?"

TASTY TRIANGLE
TALK ABOUT S.T.E.A.M.

What other circles can be found in the paragraph?
once, upon, Rhombi, lovced, cool, word, also, out, how, to, for, lot, of, fun

How might Rhombi test a shape to see that it is a square?
Look for 2 right angle corners. Measure all sides: they should be equal.

What other tirangles can be seen around the classroom? **Answers will vary.**

What shapes is the origami paper? **square**

What does an accordian look like? **zig-zag fold**

What were the mice trying to haul? **cereal box**

Where do you think the mice were taking their treasure? **home**

Estimate the weight of a newspaper. **12-16 ounces**

What day is it in the Rhombi story? What is today's actual date? **Wednesday, Vary**

Why was Rhombi surprised when she returned to the park bench?
Her accordian folded paper was under a newspaper.

What shape did Rhombi see when she bent to examine the paper? **triangle**

Where else did Rhombi see triangles in the park? **bench, bridge, bicycle**

Where would triangles be found in a bicycle? **at least 2 triangles hold the frame**

Why did Rhombi wish the mice had a bridge?
Mice could drag their box across the bridge.

What materials did Rhombi have to use? **Origami paper**

Who might help Rhombi? **builders**

How would you fold a paper bridge? **zig zag fold the paper like an accordian**

UNIT 3 CHALLENGE: TASTY TRIANGLE

"Designers, can you fold a paper bridge that will help the mice get their box home?"

1. The bridge will need to use triangles in its design.
2. Test many types of paper to find the strongest for your design.
3. The bridge should support at least 6 ounces (material and tray).
4. To test papers, rest a tray on the "bridge" with materials on tray.

List all the different types of paper you can remember.

Draw your bridge design.

Draw each of the pieces you will need to complete the challenge.

paper?

scale?

tray?

what else?

On the back of this paper, write out the steps you will need to take.

UNIT 3 RUBRIC: TASTY TRIANGLE

TASK **Students will use their knowledge of triangles to troubleshoot and build a folded paper bridge that holds at least 6 ounces of material including the tray on which it rests.**

	Content	Organization	Presentation
1	• Is well thought out and supports the solution to the challenge or question • Reflects application of critical thinking • Has clear goal that is related to the topic • Is accurate	• Information is clearly focused in an organized and thoughtful manner • Information is constructed in a logical pattern to support the solution	• Format enhances the content
2	• Supports the solution • Has application of critical thinking that is apparent • Has no clear goal • Has some factual errors or inconsistencies (i.e. wheels do not rotate around axle)	• Project has a focus but might stray from it at times (more concerned with form than function) • Information appears to have a pattern, but the pattern is not consistently carried out in the project	• Format does not suit the content
3	• Provides inconsistent information for solution • Has no apparent application of critical thinking • Has no clear goal • Has significant factual errors, misconceptions, or misinterpretations	• Content is unfocused and haphazard • Information does not support the solution to the challenge or question • Information has no apparent pattern	• Presentation appears sloppy and/or unfinished

TRIANGLE

ABOUT THE TRIANGLE

- Triangles are polygons with the least possible number of sides (three).
- The three internal angles of a triangle always add to 180 degrees.
- An equilateral triangle has three sides of equal length and three equal angles.
- The longest side of a right angle triangle is called the hypotenuse, it is always found opposite the right angle.
- Trigonometry is the study of the relationship between the angles of triangles and their sides.
- Triangle shapes are often used in construction because of their great strength.

UNIT 4

FLYING FIVES

Essential Questions

1. How many edges and vertices are present in a regular pentagon?
2. What important monument is built in the shape of a pentagon?
3. How would you draw a pentagon?
4. What are you trying to accomplish in Rhombi's Challenge?

FLYING FIVES

How many different objects, flora and fauna exist in a pentagonal shaped piece of land? What inventions would help the class? Design and improve signage material. Learn to read infographics. Investigate and create comic strips with 5 panels. Discover the true size of America's famous pentagon.

Students will count to 100 and bundle, compose and decompose numbers. Halves, wholes, and quarters are introduced in folding paper to make pentagons. Discover symmetry and deepen measurement and geometric understanding.

Help Rhombi drop hiking supplies into the cave for a spelunking trip.

ENDURING UNDERSTANDING

Two dimensional shapes create new shapes.

The length of an object may be discussed using whole numbers.

Organizing information helps in discussion.

Number information may be organized for the purpose of answering questions.

Shapes may be identified by their attributes.

Symmetry means a shape can be folded so edges will match

Asking questions and taking part in discussions increases understanding of a subject.

FLYING FIVES
CURRICULUM CONNECTIONS

Rhombi needs a pentagonal parachute.

MATH BACKGROUND - Pentagons

A pentagon is a 5 sided polygon with interior angles that add to 540 degrees. Regular pentagons have edges/sides of equal length and interior angles / vertices of 108 degrees. The edible plant okra is shaped like a pentagon.

TECHNOLOGY CONNECTIONS

Julett Morris Elementary Explains Shapes
http://www.schooltube.com/organization/122887/

Students use shapes to build.
http://bit.ly/q8OY97

Shapes Song by Peter Weatherall
http://www.youtube.com/user/peterweatherall/videos

HISTORICAL CONNECTIONS

The Pentagon is home to Department of Defense and has three times the floor space of the Empire State Building in New York. According to the website, "Despite 17.5 miles of corridors it takes only seven minutes to walk between any two points in the building." Pentagon Tours explain that there "are 19 escalators, 131 stairways, 284 restrooms, 691 water fountains and enough telephone wire in the building to wrap around the world four and one-half times!"

ART CONNECTIONS

Draw a pentagon with pencil and ruler.

Use the straight edge to draw a line from corner to corner diagonally. Repeat in the opposite corners. Draw a dot in the center.

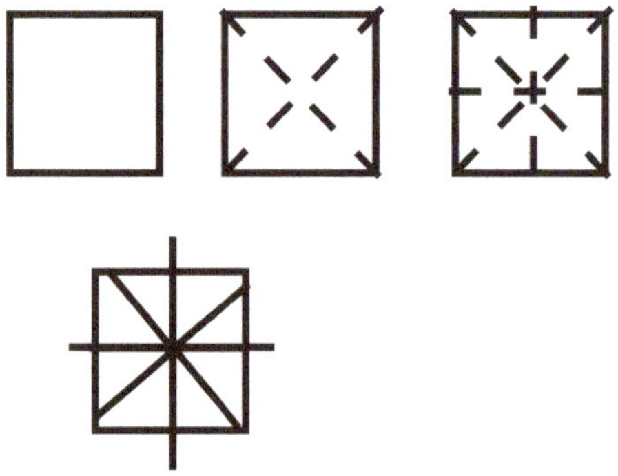

READING CONNECTIONS

Introduce each activity and lesson with a book. Books are listed on the Read about STEAM page. Jot additional lists of books from your favorites here.

1.

2.

3.

4.

FLYING FIVES
CURRICULUM CONNECTIONS

COMMON CORE CONNECTIONS

ELA/Literacy
W.K.7 Participate in shared research and writing projects (e.g., explore books by a favorite author and express opinions). (K-ESS2-1) add venn

Mathematics
MP.2 Reason abstractly and quantitatively. (K-ESS2-1)
MP.4 Model with mathematics. (K-ESS2-1)
K.CC.A Know number names and the count sequence. (K-ESS2-1)
K.MD.A.1 Describe measurable attributes of objects, such as length or weight. Describe several measurable attributes of a single object. (K-ESS2-1)
K.MD.B.3 Classify objects into given categories; count the number of objects in each category and sort the categories by count. (K-ESS2-1)

NEXT GEN SCI CONNECTIONS

K-2-ETS1-1. Ask questions, make observations, and gather information about a situation people want to change to define a simple problem that can be solved through the development of a new or improved object or tool.
K-2-ETS1-2. Develop a simple sketch, drawing, or physical model to illustrate how the shape of an object helps it function as needed to solve a given problem.
K-2-ETS1-3. Analyze data from tests of two objects designed to solve the same problem to compare the strengths and weaknesses of how each performs.

RESOURCES FOR UNIT 4: S.T.E.A.M. ACTIVITIES FOR "FLYING FIVES"

Rhombi Audio Download and Videos Available April 2014

UNIT 4: FlyingFives: Science: What will you discover in a small piece of land?
UNIT 4: FlyingFives: Technology: Can students design a new tool?
UNIT 4: FlyingFives: Engineering: Design and improve signage material.
UNIT 4: FlyingFives: Art & Music: Investigate and create comic strips with 5 panels.
UNIT 4: FlyingFives: Mathematics: Discover the true size of America's famous pentagon.

UNIT 4: FlyingFives: ELA Activity: Read Rhombi's "FLYING FIVES." Student's illustrate the story.

FLYING FIVES
STEAM BOOKSHELF

SCIENCE

Grand Canyon
by Cari Meister

One Small Square (caves)
by Donald Silver

Shapes in Sports
by Rebecca Rissman

TECHNOLOGY

Goodnight, Goodnight Construction Site
Sherri Duskey Rinker

Tool, Time, Twist: A Brief History of Tools Through Time
by David Shapiro

http://www.trafficsign.us

ENGINEERING

Monkey with a Tool Belt
by Chris Monroe

The Kids' Building Workshop
by Craig Robertson

Shapes in Transportation
by Jennifer Rozines/ Roy Roy

ART & MUSIC

13 Art Inventions Children Should Know
by Florian Heine

Shapes (Museum of Modern Art)
by Philip Yenawine

Math-terpieces
by Greg Tang

MATH

Pentagon
by Sheila Rivera

How Long or How Wide?: A Measuring Guide
by Brian P. Cleary

ELA

To Root to Toot to Parachute: What Is a Verb
by Brian P. Cleary

How Margaret E. Knight Became an Inventor
by Marvelous Mattie
Emily Arnold McCully

STEMVESTIGATIONS: STAKE OUT

SCIENCE

READ **We're Hunting for a Pentagon** *page 44*

DISCUSS Explore the twilight world of caves with bats, crickets, millipedes, blind cave fish, and salamanders.

EXPLAIN Caves are habitats that support many species. Because they are not exposed to direct sunlight, the plants and animals that live in caves have different characteristics than those living in the sunlight.

MOVE Play "Going on a Cave Hunt." In the spirit of "Going on a Bear Hunt," sit in a circle with legs folded. Slap legs in a rhythm (slap, slap, clap, clap, slap, slap, clap, clap). Each person in the circle adds a scene to the chant. For a sample, visit Schooltube for videos.
http://bit.ly/rtpbNU or
www.sophie-world.com/activities/singalong-were-going-bear-hunt

INVESTIGATE

What will you find in one small shape of your school yard? Mark off pentagonal areas with sports tape or ribbon around stakes in the ground. No stakes handy? Use pencils in the ground. Use the pentagon pattern in Art Connections to create staked out shapes for examination. For 10 minutes, carefully observe and collect a list of objects (and their shapes) found in each pentagon of earth.

ELA Present findings to the class.

STEMVESTIGATIONS: TOOLS

TECHNOLOGY

READ Tool Time *page 46*

DISCUSS A tools are nonconsumable physical objects used to achieve a goal. Tool use has been documented since the beginning of human history, and animals also employ tools.

EXPLAIN Humans have created tools to simplify work since the stone age. Explore the history of tools from the stone hand ax to the modern toolbox. 20 tools and 2.5 million years of tool making offer a lot to think about.

MOVE Visit the web to locate great songs about construction and tools. Several fun songs that can be acted out while sung.
http://www.prekfun.com
http://www.learninggamesforkids.com/science_songs.html

INVESTIGATE

What activities could be enhanced by the use of an appropriate tool? Students should brainstorm a list of things they do each day that could be easier. What tasks are time-consuming, laborious, physically difficult, or just plain annoying? Students should break into teams of four to design and develop a tool to achieve a goal. Identify the need. Explain the solution. Compare the idea to other existing tools. List materials to be used in the tool's creation. Gather supplies to create the tools and demonstrate their uses.

ELA Create "advertisements" to sell tool ideas to investors.

FLYING FIVES
STEMVESTIGATIONS: PENTAGON PATTERNS

ENGINEERING

READ Creative Design *page 48*

DISCUSS "Pentagons" introduces shapes in a repetitive format that makes vocabulary memorable. Pentagons are most widely recognized by students in the national military building and in school crossing street signage.

EXPLAIN The pentagon is a shape with 5 sides / edges and 5 corners / vertices. A regular pentagon has edges of the equal length and vertices of equal angle.

MOVE Take a walk out to the "crosswalk" sign.

INVESTIGATE

Make patterns for pentagons.
Copy and print the Polygon Pentagon for each student.

Challenge students to come up with ways to make a pattern for the pentagon. Can triangles or squares be used? Will folding help?

Hint: Slip each printed pentagon into a clear cover, or laminate the pages so students keep a clean re-usable shape.

ELA Design other school safety messages for a PENTAGONAL crosswalk sign!

STEMVESTIGATIONS: STORY IN 5 BOXES

The ARTS

READ Design Time *page 50*

DISCUSS A humanitarian is someone who designs and creates and innovates to make life better for mankind. One of his inventions was the geodesic dome. Domes make cheap, strong houses. People are so used to the box house that his invention never became popular.

EXPLAIN Inventions take many forms. The Utah State Intellectual Properties Services defines an invention as "the discovery or creation of a new material (either a new manufactured product or a new composition or matter), a new process, a new use for an existing material, or any improvements of any of these."

MOVE

INVESTIGATE

Katzenjammer Kids was the first comic strip ever published. It was printed in the New York Journal in 1897. What are students' favorite cartoons and comic strips? Cut out cartoons from the newspaper to share. Students read cartoons (or have them read) to get ideas. Use the comic strip format to write about a cool thing students would invent if they had all the time, money, resources, and abilities they could imagine.

ELA Publish a classroom "graphic novel" or comics page. Include the page in a full-fledged parent newspaper.

STEMVESTIGATIONS: THE PENTAGON

MATHEMATICS

READ The Pentagon, USA *page 52*

DISCUSS The Pentagon is one of America's best known buildings. Its design was chosen to fit an unusually shaped piece of land. When the site changed, architects and engineers kept the pentagonal shape of the building for its new location.

EXPLAIN According to GreatBuildings.com, each outer edge has a length of 921feet. Other numbers include:

- Floor area of 6.5 million square feet or 34 acres
- 3.7 million square feet are used for offices
- 80 million dollar construction cost in 1943 dollars
- 17.5 miles of corridors

MOVE Each edge of the Pentagon measures 921 feet. Walk 921feet. Walk it 5 times to simulate walking the perimeter of this huge building.

INVESTIGATE

As a class, create a scale model of the Pentagon building with blocks and other classroom construction materials. If each outer wall measures 921 feet, adjust your classroom or playground structure to measure a scaled down 9.21 feet on each edge. How will the class make sure the structure is actually a regular pentagon with equal angles? (Hint: They can print and cut out a paper sized pentagon. Lay the paper at your large pentagon's center. Assemble large walls parallel to the small pentagon's edges.) What materials will you use?

DOT TO DOT

MATERIALS

You will need an overhead projector or smart board setup. Copy the dot paper on a transparency. You will need appropriate marking tools (transparency markers or smart mouse setup).

SETUP

Project dot paper on a screen or whiteboard.

INSTRUCT THE PLAYERS

Taking turns, the first player connects two dots with either a horizontal or vertical line. The other players take their turns doing the same. Their lines may touch other player's lines, but they don't have to. When a player draws the fourth line to create a box, the team claims the box and writes the first letter of its team name inside the box.

STUDENTS BREAK INTO TWO CLASS TEAMS OR SMALL TEAMS

Play. Once all of the possible horizontal and vertical lines in the grid have been turned into boxes, each team counts the number of boxes it owns. The team with the most boxes wins.

UNIT 4

FLYING FIVES

Once upon a time there was Rhombi. Rhombi loved shapes and found them everywhere. Her friends had helped her design a cart to carry things. She had a handy tool for checking a square's shape. She even had a plan for building strong bridges.

Rhombi was sitting on a very strong bench outside her school making the list for a camping trip. Her team was going spelunking soon and needed to drop a bunch of supplies down into a cavern. They only had enough rope for themselves, and the supplies were really heavy.

Rhombi waved to some friends and noticed the yellow school crossing sign by the road. The sign had 5 sides. She'd never really noticed it before. She opened a search engine on her phone and typed "school crossing shape." The shape of the sign was called a regular pentagon.

Next, Rhombi typed “five-sided objects.” The search engine listed other objects in the shape of regular pentagons. Rhombi always checked facts from the internet, so she decided to look for pentagons on her own.

FLYING FIVES

Rhombi looked at a morning glory in the garden. While she was examining the surface of a soccer ball, she caught sight of motion in the sky. Rhombi turned to see a skydiving team land in the nearby field. Their fabric parachutes slowed a skydivers fall to Earth from great heights. Suddenly, Rhombi had an idea.

If you could drop people, could you also drop supplies? Maybe her spelunking problem could be solved with pentagons. Could she make a five-sided parachute for her supply drop? Rhombi needed help. She looked out from the pages of her world and asked, "Designers, will you help me create a pentagonal parachute?"

FLYING FIVES
TALK ABOUT S.T.E.A.M.

What shape helped Rhombi's cart roll? circle

How did Rhombi check a square's shape? She used the corner of a notebook paper.

What shape made her bridge strong? triangle

What shape made her bench strong?triangle

What is spelunking? cave diving

Why couldn't Rhombi drop her supplies by rope? The rope was too short.

What color is the "school crossing" sign? yellow

How many sides are present on the sign? 5

What is the shape of the 5-sided sign? pentagon

What did Rhombi use to find her list of pentagonal objects? internet search engine

Name at least 3 objects that have regular pentagons? star fruit cross section, crosswalk sign, Pentagon building in Washington, D.C., USA

What is a starfruit? green tropical fruit

What are the rules for a regular pentagon? 5 edges of equal length with 5 vertices of equal angle

What was Rhombi's spelunking problem? She needed to get supplies to the cave floor.

What is a parachute? Flexible fabric "umbrella" that uses drag to slow an object's rate of fall.

RHOMBI'S ADVENTURES IN 2D
UNIT 4 CHALLENGE: FLYING FIVES

"Designers, will you create a pentagonal parachute to drop supplies safely to the cave floor?"

1. For your scale model, supplies should weigh about 3 ounces.
2. The parachute will need to hold carry a 3 ounce load and stay in the air for more than 1.5 seconds before landing.
3. Get creative with materials. What makes the best parachute?

Draw a pentagonal parachute.

List at least 3 different parachute materials that you would like to test.

Draw or list each of the pieces you will need to complete the challenge.

string?
fishing line?
3 fabrics?
supplies?
pentagon pattern?

On the back of this paper, write out the steps you will need

RHOMBI'S ADVENTURES IN 2D
UNIT 4 RUBRIC: FLYING FIVES

TASK **Students will use their knowledge of pentagons to troubleshoot and build a parachute to safely drop supplies.**

	Content	Organization	Presentation
1	• Is well thought out and supports the solution to the challenge or question • Reflects application of critical thinking • Has clear goal that is related to the topic • Is accurate	• Information is clearly focused in an organized and thoughtful manner • Information is constructed in a logical pattern to support the solution	• Format enhances the content
2	• Supports the solution • Has application of critical thinking that is apparent • Has no clear goal • Has some factual errors or inconsistencies (i.e. wheels do not rotate around axle)	• Project has a focus but might stray from it at times (more concerned with form than function) • Information appears to have a pattern, but the pattern is not consistently carried out in the project	• Format does not suit the content
3	• Provides inconsistent information for solution • Has no apparent application of critical thinking • Has no clear goal • Has significant factual errors, misconceptions, or misinterpretations	• Content is unfocused and haphazard • Information does not support the solution to the challenge or question • Information has no apparent pattern	• Presentation appears sloppy and/or unfinished

RHOMBI'S POLYGON ROUND-UP
PENTAGON

ABOUT THE PENTAGON

- A pentagon is a 5 sided polygon with interior angles that add to 540 degrees.
- Regular pentagons have sides of equal length and interior angles of 108 degrees.
- The US Department of Defense headquarters is named 'the Pentagon'.
- The edible plant okra is shaped like a pentagon.

ON THE MOVE

ON THE MOVE

Essential Questions

1. What tools should be used to measure ounces & feet?
2. What units should be used to measure mass/weight and distance/length?
3. Can you talk about shapes using attributes?
4. What will happen to the shapes in your design when they are transformed?

ENDURING UNDERSTANDING

Numbers can represent quantity, position, location, & relationships.

Computation involves taking apart and bundling numbers

Objects may be described by their attributes.

Points, lines, and planes are the basis for geometry.

Transformation of an object does not change its shape.

Objects have distinct attributes that can be measured.

Standard units provide common anchors for discussion.

While this project may be approached as a stand-alone investigation, it also acts as an authentic assessment. Students will apply knowledge of physical science, earth science, materials, attributes, geometry, numeracy and measurement to design and build a scale model.

Rhombi and her friends are headed out for a camping trip. Rhombi will have many adventures, cross rivers, drop supplies, and need help packing her gear. Students will work in teams to troubleshoot her trip.

Logbooks should contain drawings, designs, and "notes" about the project.

ON THE MOVE

Once upon a time, there was Rhombi. Rhombi loved shapes, and she found them everywhere. Triangles showed up in benches and bridges. Squares were found in tiles on the floor. Circles helped people move from place to place. Even pentagons could be found in morning glories and parachutes!

Rhombi was restless. School was out for the summer, and Rhombi wanted a change. She decided to pack her things for a road trip. Rhombi did not have a lot of money, so she would have to plan carefully.

Rhombi would need food, water, clothes, camping gear and a way to move all that stuff. She checked the map to decide on a destination. Where did she want to go? The national park looked great. It would be cold at night, but it had great hikes.

Rhombi decided to head west toward the mountains and lakes. She drew her route on the map and chose a design for her vehicle. Designers had helped her, and she had a lot of plans from which to choose.

Rhombi made her packing list for a 5 day trip. She included the food, water, clothing, blankets and tent she would need to stay warm, fed, hydrated and dry.

Rhombi knew that she would need to cross the gap where a bridge had broken. That meant she'd need to carry materials for crossing in her vehicle. With all the designs given to her, she would need to choose the best bridge.

Rhombi would need to drop her supplies from a great height before hiking down into a river gorge. The trail was too steep for a rolling wagon, and she couldn't carry all that stuff herself. Which parachute design would she use?

Finally, once she was at the river's edge, Rhombi would need to get her supplies across the water. Hmmm... Would she use a boat or a bridge or a flying contraption? How could she move all her things across the river?

Rhombi would need some help with all these decisions. She looked out beyond the pages of her world and asked, "Designers, will you work in teams to help me travel the trail, drop into the canyon, and cross a river with supplies?"

ON THE MOVE
TALK ABOUT S.T.E.A.M.

What is the season? summer

Rhombi must plan carefully. Why? She has limited funds.

What would Rhombi need for her trip?
food, water, clothing, shelter

Where would you go if you could take a trip?
Answers will vary.

In which direction will Rhombi head?
West

On what will Rhombi draw her route? map

What kind of vehicle will Rhombi use? rolling cart or wagon? Answers will vary.

How long will Rhombi be on the road? 5 days

What important things will Rhombi need for survival and comfort? food, water, shelter, clothing, vehicle

What broke? bridge

What would Rhombi need to drop? supplies

Why couldn't Rhombi just roll her supplies down the path? The path is too steep.

Which of the parachute designs should Rhombi choose? Answers will vary but include pentagon.

What will Rhombi need to cross? river

What are all the different ways Rhombi could cross the river?Answers will vary.

What shapes would be most useful in helping Rhombi cross? Answers will vary.

EXTENDING THE ADVENTURE
ON THE MOVE CHALLENGE: PLAN, PACK, TRAVEL

“Designers, will you help me pack for my trip and carry my gear? Some teams will design my boat. Some will create the way to drop supplies into a canyon. Some teams will design a wagon or cart that helps me move supplies and camping gear.”

EXTENSION CHALLENGE RUBRIC: ON THE MOVE

TASK **Students will use their knowledge of polygons to troubleshoot Rhombi's planned camping trip.**

	Content	Organization	Design
1	• Is well thought out and supports the solution to the challenge or question • Reflects application of critical thinking • Has clear goal that is related to the topic • Is accurate	• Information is clearly focused in an organized and thoughtful manner • Information is constructed in a logical pattern to support the solution	• Format enhances the content • Project is well-constructed. • Project meets design specifications.
2	• Supports the solution • Has application of critical thinking that is apparent • Has no clear goal • Has some factual errors or inconsistencies (i.e. wheels do not rotate around axle)	• Project has a focus but might stray from it at times (more concerned with form than function) • Information appears to have a pattern, but the pattern is not consistently carried out in the project	• Format does not suit the content. • Project is not well-constructed but does show evidence of effort. • Project does not perform well.
3	• Provides inconsistent information for solution • Has no apparent application of critical thinking • Has no clear goal • Has significant factual errors, misconceptions, or misinterpretations	• Content is unfocused and haphazard • Information does not support the solution to the challenge or question • Information has no apparent pattern	• Presentation appears sloppy and/or unfinished. • Project is sloppy. • Project cannot perform the designated task at all.

APPENDIX

POLYGONS

POLYGON NET PATTERNS 123 - 124

POLYGON TRACING SHAPES 125 - 133

STEMVESTIGATIONS ACTIVITY PAGES

ROUND AND ROUND 136 - 139
MOON PHASES, ROVERS, WHEELS, QUILTS, RADIUS

LOOKING FOR SQUARES 140 - 145
GROWING THINGS, PRESENTATIONS, SHAPE WALK, QUILT PATTERNS, PAPER SQUARES

TASTY TRIANGLES 146 - 150
GROWING POLYGONS, PRESNTATIONS, MATERIALS, TRI ART, ANGLES

FLYING FIVES 151 - 155
STAKE OUT, TOOLS, MATERIALS, 5 BOX STORIES, THE PENTAGON

MINDBUGS IN MEASUREMENT ACTIVITY PAGES

INCH BY INCH 157 THINK ABOUT TIME 158 USING THE DIGITAL SCALE 159 COORDINATE PLANES 160
USING THE STOPWATCH 151 STANDARDIZING MEASUREMENT 162

CORRELATIONS AND ASSESSMENTS

VOCABULARY 163

PRE AND POST ASSESSMENT 164 - 172

PLANNING PAGES 173 - 175

NEXT GENERATION SCIENCE STANDARDS CORRELATIONS CHART 174 - 175

COMMON CORE CHECKLISTS 176 -181

WORD WALL VOCABULARY FOR MATH AND SCIENCE 182 - 185

REFERENCES AND WORKS CITED 188

CUBE NET

INSTRUCTIONS: Print this net for use in a center or during STEMVESTIGATIONS.

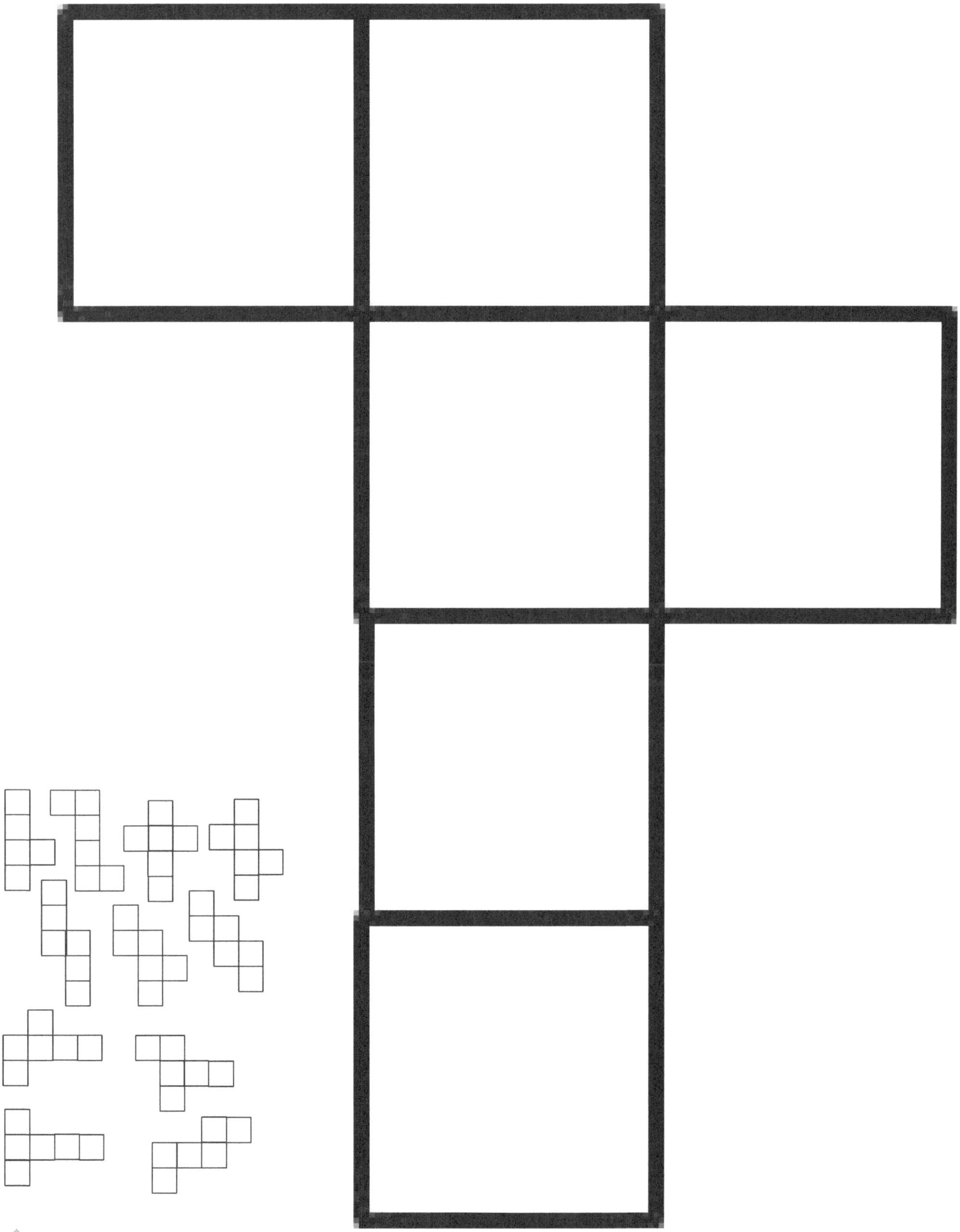

APPENDIX
PYRAMID NET

INSTRUCTIONS: Print this net for use in a center or during STEMVESTIGATIONS.

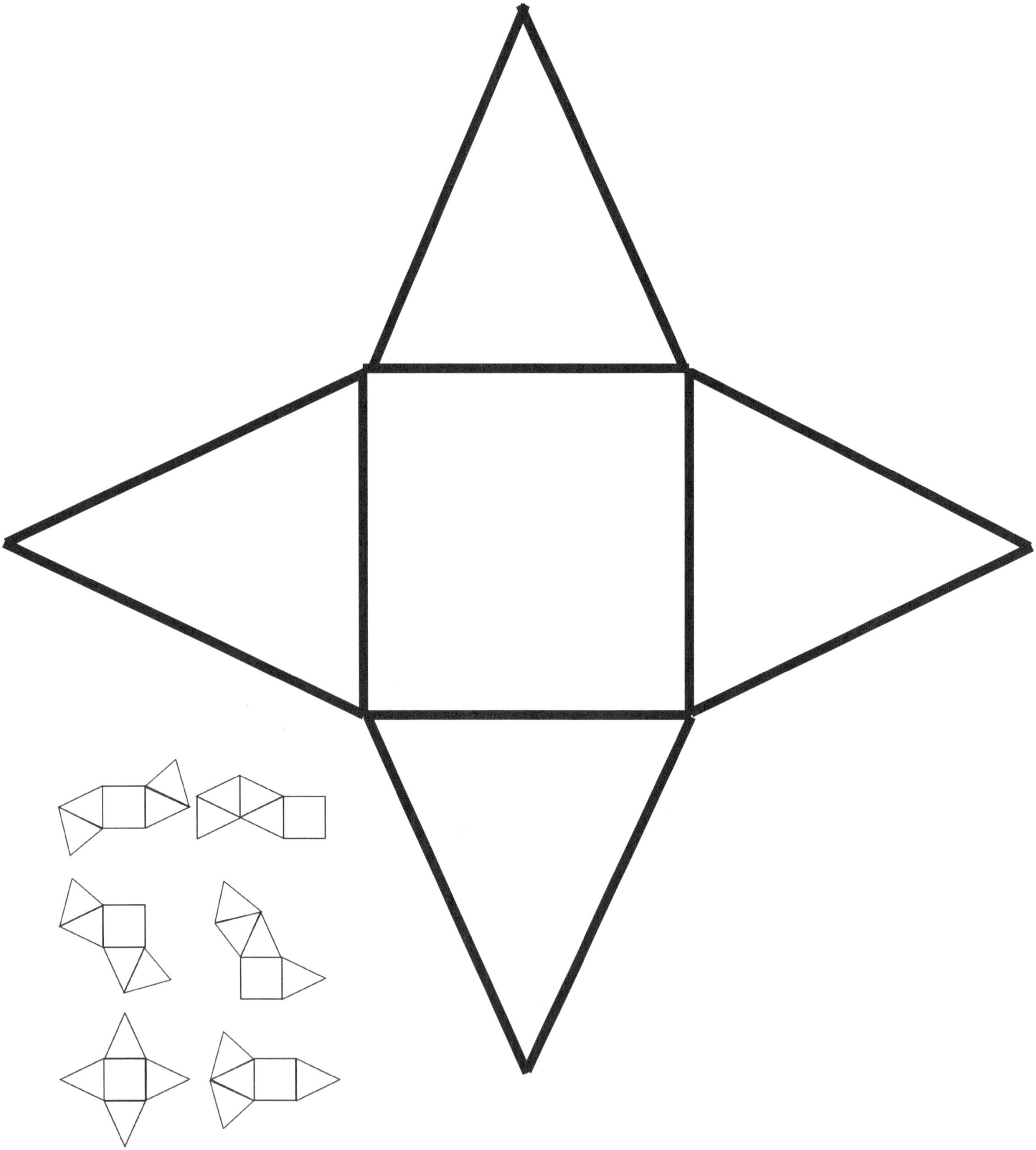

RHOMBI'S POLYGON ROUND-UP
CIRCLE

ABOUT THE CIRCLE

- In the math world, circle refers to the boundary of the shape. "Disk" is used to refer to the whole shape, including the inside.
- A straight line from the center of a circle to the edge is called the radius.
- A straight line that passes from one side of a circle to the other through the center is called the diameter.
- The distance around the outside of a circle is called the circumference.All points on the edge of a circle are the same distance to the center.
- The value of Pi (π) to 2 decimal places is 3.14, it comes in handy when working out the circumference and area of a circle.
- Circles have a high level of symmetry.

RHOMBI'S POLYGON ROUND-UP
TRIANGLE

ABOUT THE TRIANGLE

- Triangles are polygons with the least possible number of sides (three).
- The three internal angles of a triangle always add to 180 degrees.
- An equilateral triangle has three sides of equal length and three equal angles.
- The longest side of a right angle triangle is called the hypotenuse, it is always found opposite the right angle.
- Trigonometry is the study of the relationship between the angles of triangles and their sides.
- Triangle shapes are often used in construction because of their great strength.

RHOMBI'S POLYGON ROUND-UP
RECTANGLE

ABOUT THE SQUARE

- A square is a polygon with 4 sides of equal length and 4 right angle corners (90 degree corners).
- Because it has 4 sides of equal length, a square is a regular quadrilateral.
- A square is also a rectangle with equal sides and a rhombus with right angles.
- The perimeter of a square is 4 times the length of one side.
- Opposite sides of a square are parallel.
- The internal angles of a square add to 360 degrees.
- A square has 4 lines of reflectional symmetry.

RHOMBI'S POLYGON ROUND-UP
PENTAGON

ABOUT THE PENTAGON

- A pentagon is a 5 sided polygon with interior angles that add to 540 degrees.
- Regular pentagons have sides of equal length and interior angles of 108 degrees.
- The US Department of Defense headquarters is named 'the Pentagon'.
- The edible plant okra is shaped like a pentagon.

HEXAGON

ABOUT THE HEXAGON

- A hexagon is a 6 sided polygon with interior angles that add to 720 degrees.
- Regular hexagons have sides of equal length and interior angles of 120 degrees.
- Beehive cells are hexagonal.

RHOMBI'S POLYGON ROUND-UP
HEPTAGON

ABOUT THE HEPTAGON

- A heptagon is a 7 sided polygon with interior angles that add to 900 degrees.
- Regular heptagons have sides of equal length and interior angles of 128.57 degrees.
- The British 50 and 20 pence coins are curved heptagons.

RHOMBI'S POLYGON ROUND-UP
OCTAGON

ABOUT THE OCTAGON

- An octagon is an 8 sided polygon with interior angles that add to 1080 degrees.
- Regular octagons have sides of equal length and interior angles of 135 degrees.

RHOMBI'S POLYGON ROUND-UP
NONAGON

ABOUT THE NONAGON

- A nonagon is a 9 sided polygon with interior angles that add to 1260 degrees.
- Regular nonagons have sides of equal length and interior angles of 140 degrees.

RHOMBI'S POLYGON ROUND-UP
DECAGON

ABOUT THE DECAGON

- A decagon is a 10 sided polygon with interior angles that add to 1440 degrees.
- Regular decagons have sides of equal length and interior angles of 144 degrees.

RHOMBI'S POLYGON ROUND-UP
TEACHER'S BACKGROUND

GENERAL INFORMATION ON POLYGONS

A polygon is a geometric figure in two dimensions with three or more sides. The name comes from two Greek words, poly, meaning "many," and gon, meaning "angle." A polygon always has as many angles as it has sides. Polygons are named to indicate the number of sides or angles they contain. Thus, a hexagon has six (hexa- means "six") sides and six angles. The sum of a polygon's exterior angles is 360°...
To find the internal angles of a polygon use this formula:
Multiply 180 by (the number of sides minus 2). A hexagon has 6 sides. $6 - 2 = 4$ Hexagon is $180° \times 4 = 720°$

SUM of INTERNAL ANGLES

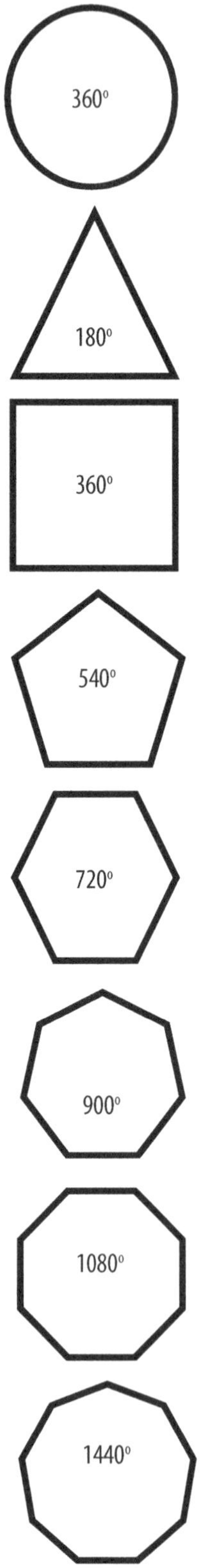

POLYGON TERMINOLOGY
Parts and properties of polygons.

Side: Any one of the straight lines that make up the polygon.

Vertex: A point where any two of the sides of a polygon meet to form an angle.

Angle: A figure formed by the intersection of two sides.

Diagonal: A line that joins any two nonadjacent (not next to each other) vertices.

Perimeter: The sum of the length of all sides.

Area: The space enclosed within the polygon.

TYPES OF POLYGONS

Equilateral: A polygon in which all sides are equal in length.

Equiangular: A polygon in which all angles are the same size.

Regular: A polygon that is both equilateral and equiangular. This is the one most often used in schools.

EXAMPLES OF POLYGONS
The most common kinds of polygons include:

Parallelogram: A quadrilateral (four-sided figure) in which both pairs of sides are parallel and equal.

Rhombus: A parallelogram in which all four sides are equal.

Rectangle: A parallelogram in which all angles are right angles.

Square: A rectangle in which all four sides are equal.

RHOMBI'S ADVENTURES IN 2D STEMVESTIGATION ACTIVITY AIDS

ROUND AND ROUND

HARVEST SHAPES 136

PATTERNS 137

WHEELS 138

QUILTS 139

CIRCLES SHAPE WALK 140

ON THE SQUARE

GROWING THINGS 141

QUADRILATERALS 142

SQUARES SHAPE WALK 143

QUILT PATTERNS 144

PAPER SQUARES 145

TASTY TRIANGLES

POLYGONS 146

PRESENTATIONS 147

GROWING 148

TRIANGLES IN ART 149

WHAT'S YOUR ANGLE 150

FLYING FIVES

STAKE OUT THE PENTAGON 151

TOOLS THROUGH TIME 152

MATERIALS 153

STORY IN 5 BOXES 154

WALK THE PENTAGON 155

ROUND AND ROUND
STEMVESTIGATIONS: MOON PHASES

Create a chart of the moon's phases.

1. Observe the moon each night for one month. *How many week-long charts will you need?*
2. Sketch what you see.
3. Write the name of the moon's phase. *first quarter, full moon, third quarter, and new moon*
4. Add notes in your journal to remind yourself what you observed.

DATE	SKETCH	WEATHER	PHASE

Check the moon phases for any date at this URL. *http://stardate.org/nightsky/moon*

ROUND AND ROUND
STEMVESTIGATION: ROVERS

ACTIVITY PAGE 24

Use shapes to draw your own rover on the moon.

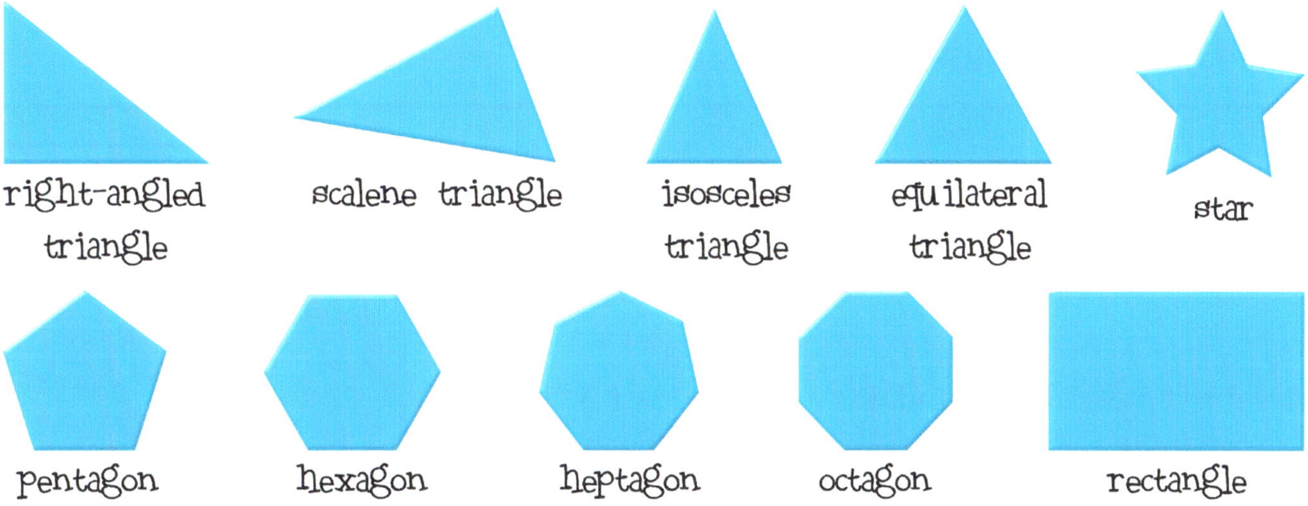

ROUND AND ROUND
STEMVESTIGATIONS: WHEELS

Move weight from START to FINISH as smoothly as possible.

Drag, roll, pull, push, or move 6 ounces of weight a distance of 7 inches.
Challenge: Move as little sand as possible.

What vehicle will your group design? Why?

TYPES OF VEHICLES
sled
wagon
sledge
boat
cart

____

____

____

Test your vehicle in the sand. Run tests at least 4 times before changing your design. Measure the trench left in the sand. What measurement tool will you use?

RUN	VEHICLE 1	DEPTH IN MILLIMETERS
1		
2		
3		
4		

Will the best result be the smallest or highest number on your data table?

RUN	VEHICLE 2	DEPTH IN MILLIMETERS
1		
2		
3		
4		

Are any of the simple machines useful in this challenge?

- Wedge
- Wheel and Axle
- Lever
- Inclined Plane
- Screw
- Pulley

ROUND AND ROUND
STEMVESTIGATIONS: PATTERN QUILTS

Write a word for each attribute that helps you plan your quilt.

ATTRIBUTES	EXAMPLE	YOUR WORDS
COLORS	*red, blue, green, pink, yellow, orange, brown, white, black, tan, purple*	
PATTERN OR PICTURE	*plaid, stripe, dots, criss-cross, scribble*	
TEXTURE	*fuzzy bumpy puffy flat smooth scratchy*	

Cut out these circles to use in your quilt.

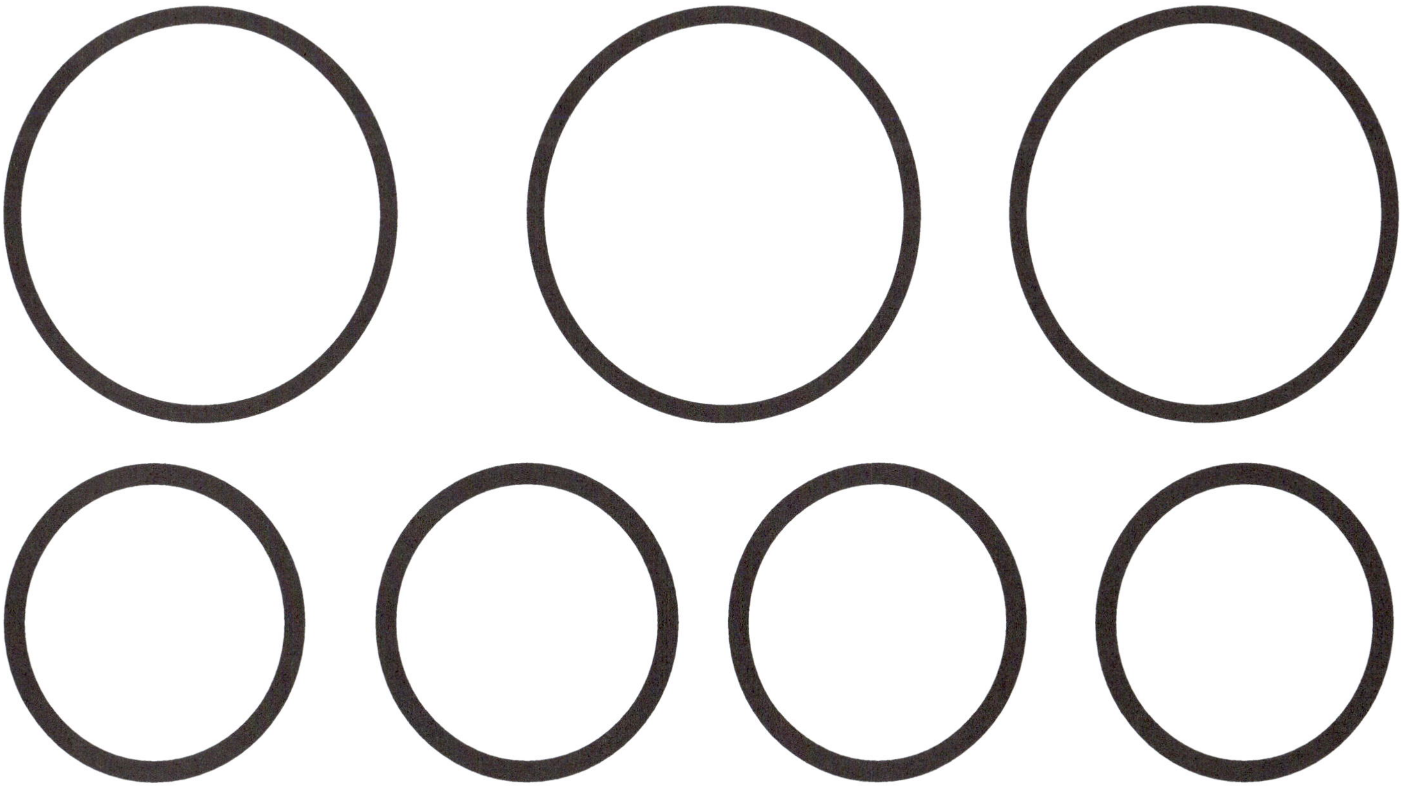

ROUND AND ROUND
STEMVESTIGATIONS: SHAPE WALK

Take a shape walk to locate circles in and around the school.

Use or draw a map of the school to plan your walk. Map a route that visits more than one area of the campus. On your map, mark the number of each shape on the map that matches your sketch.

1		2		3	
4		5		6	
7		8		9	
10		11		12	

LOOKING FOR SQUARES
STEMVESTIGATIONS: GROWING THINGS

Tally the vegetables and fruits in your lunch.

For one week, track the foods in your lunch. At the end of the week, add your data to the class chart. Compare the data recorded by day, by boy vs. girl, or find other ways to look at the information.

Day and Date	Grains	Fruits	Vegetables	Dairy	Protein
Monday					
Tuesday					
Wednesday					
Thursday					
Friday					

LOOKING FOR SQUARES
STEMVESTIGATIONS: QUADRILATERALS

Make a quadrilateral photo show to share with other students.

Use office software to create a presentation. Take pictures with a digital camera or smart phone. Save the images to a computer, and insert the pictures into a document. Take pictures of the following shapes if possible!

Organize your show before using the computer.

1. *INTRODUCE YOURSELF*	*2.* *EXPLAIN YOUR PROJECT.*	*3.* *DESCRIBE OR SHOW PICTURES OF THE SHAPES YOU HAVE CAPTURED.*
4. *PLACE ONE PICTURE PER PAGE.*	*5.* *DESCRIBE THE SHAPE, AND OFFER AS MANY FACTS AS YOU KNOW.*	*6.* *CLOSE THE PRESENTATION WITH SOME INTERESTING QUESTIONS FOR THE AUDIENCE.*

ACTIVITY PAGE 47

LOOKING FOR SQUARES
STEMVESTIGATIONS: SHAPE WALK

Take a shape walk. Look for squares. Draw or describe.

Compare squares. With a team of students, play a game of squares. Try to locate squares that no other team has found. In 30 seconds, write down as many squares as you see in your classroom. Your team gets a point for each unique square.

ROUND ONE

ROUND TWO

ROUND THREE

ROUND FOUR

ROUND	*POINTS*

LOOKING FOR SQUARES
STEMVESTIGATIONS: QUILT PATTERNS

Draw a quilt pattern. Use a combination of quadrilaterals.

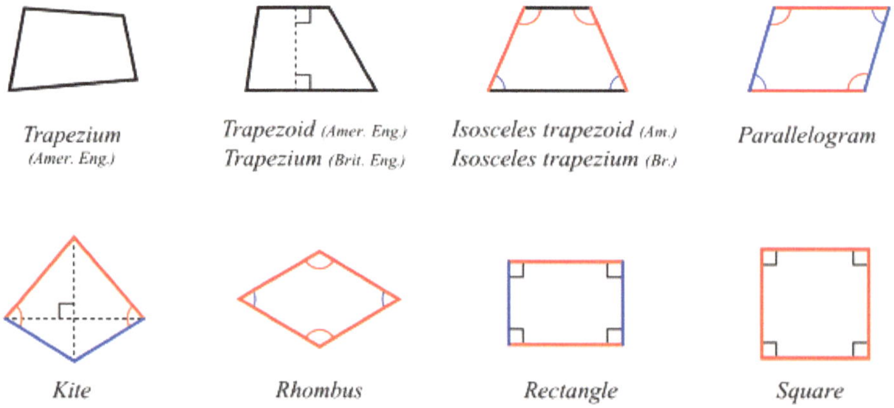

LOOKING FOR SQUARES
STEMVESTIGATIONS: PAPER SQUARES

Make your own square papers for origami. Learn to fold a shape.

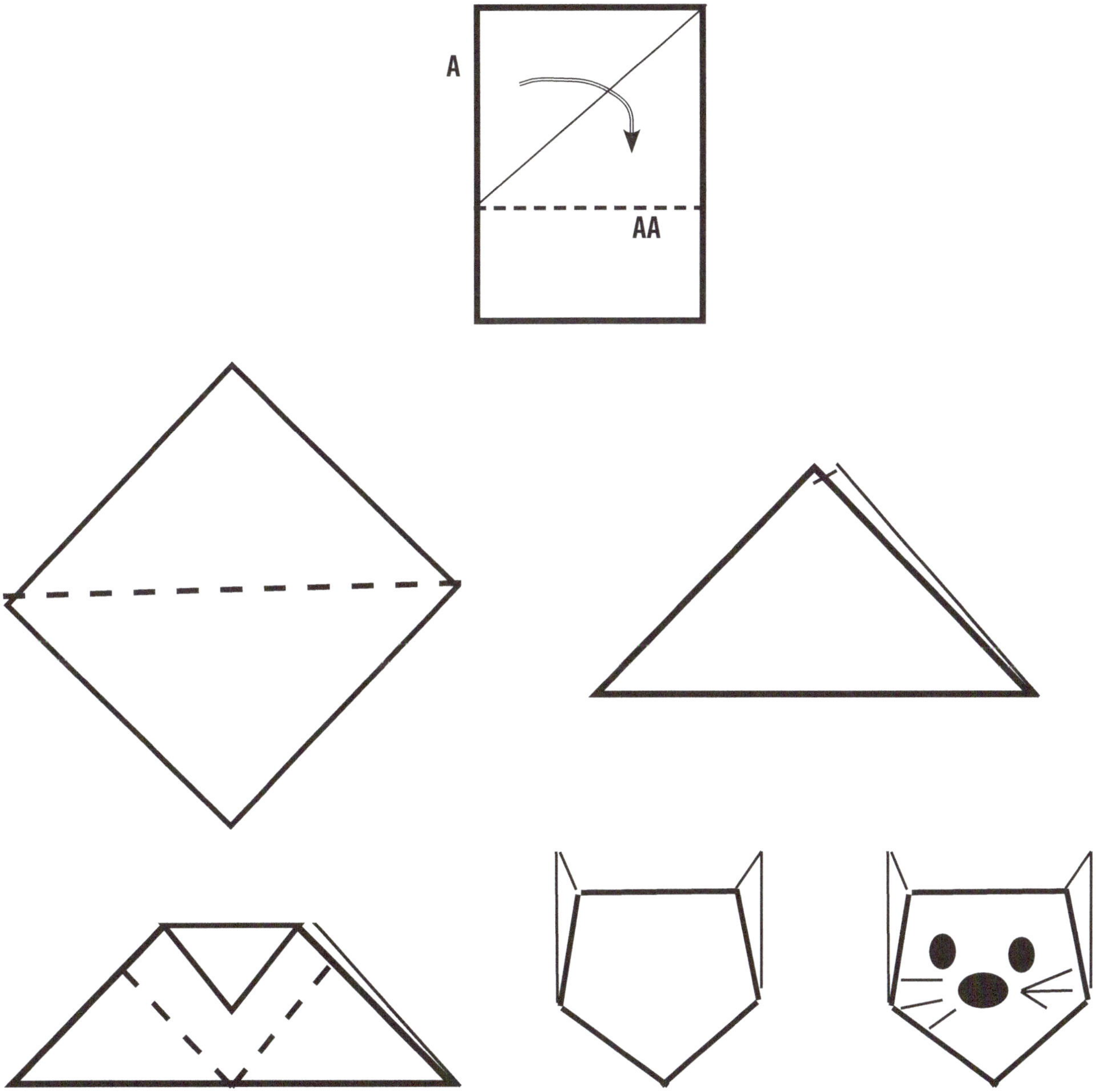

Visit the Internet for patterns and instructions.
http://www.origami-instructions.com/origami-for-kids.html

TASTY TRIANGLE
STEMVESTIGATIONS: GROWING POLYGONS

Use straws and food to make these polygons.

1. How many polygons can you name without peeking at the answers?
2. Trace each of the polygons on this page.
3. Write the name of each polygon you cannot name without peeking.

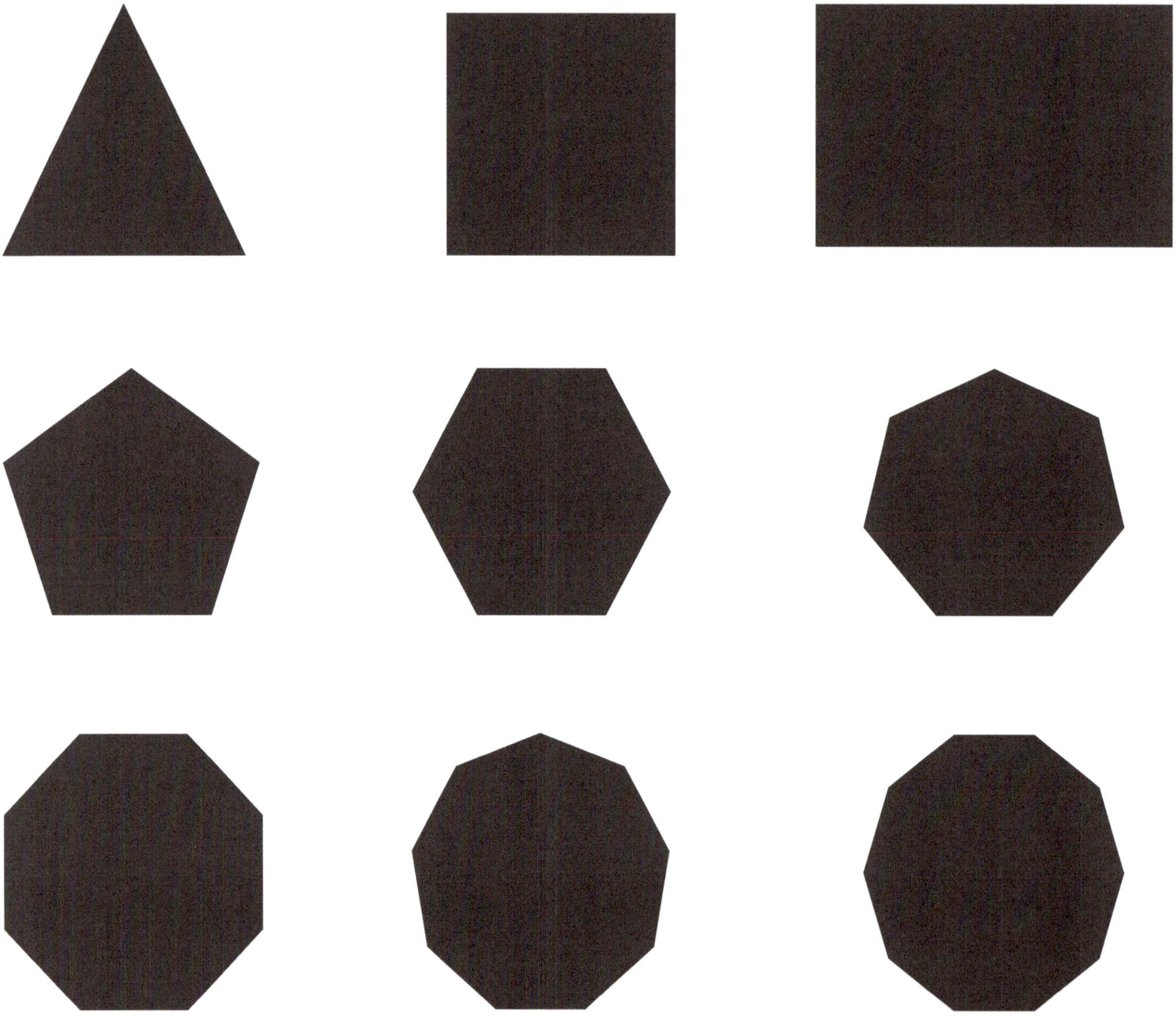

triangle, square, rectangle, pentagon, hexagon, heptagon, octagon, nonagon, decagon

STEMVESTIGATIONS: TRIANGLE PRESENTATIONS

Draw the steps for your class triangle dance!

It's time to teach your friends all about triangles and angles.

ACUTE   OBTUSE   RIGHT

Demonstrate each angle with your arms or legs or both!

Write the words for your song and dance.

Illustrate the steps for the dance.

TASTY TRIANGLE
STEMVESTIGATIONS: STRUCTURES

Make block tables. Test their ability to hold weight.

1. Test the strength of foundations and columns. Build tables from blocks and cardboard as shown in the table below.
2. Add weight to the structures until each collapses under the weight. Use a digital or balance scale to collect and record the weight at collapse.
3. What other combinations would you test?
4. How could you make the structures stronger?

Make a table.	Record weight.
1 Block and Cardboard	ounces
2 Blocks and Cardboard	ounces
3 Blocks and Cardboard	ounces
4 Blocks and Cardboard	ounces
5 Blocks and Cardboard	ounces

TASTY TRIANGLE
STEMVESTIGATIONS: TRIANGLES IN ART

Blend the three primary colors.

1. Primary colors include RED, BLUE, and YELLOW.
2. When blended, these three colors create secondary colors including ORANGE, PURPLE, and GREEN.
3. Paint the three lines as shown. Blend paint toward the center. What happens?

BLUE

YELLOW

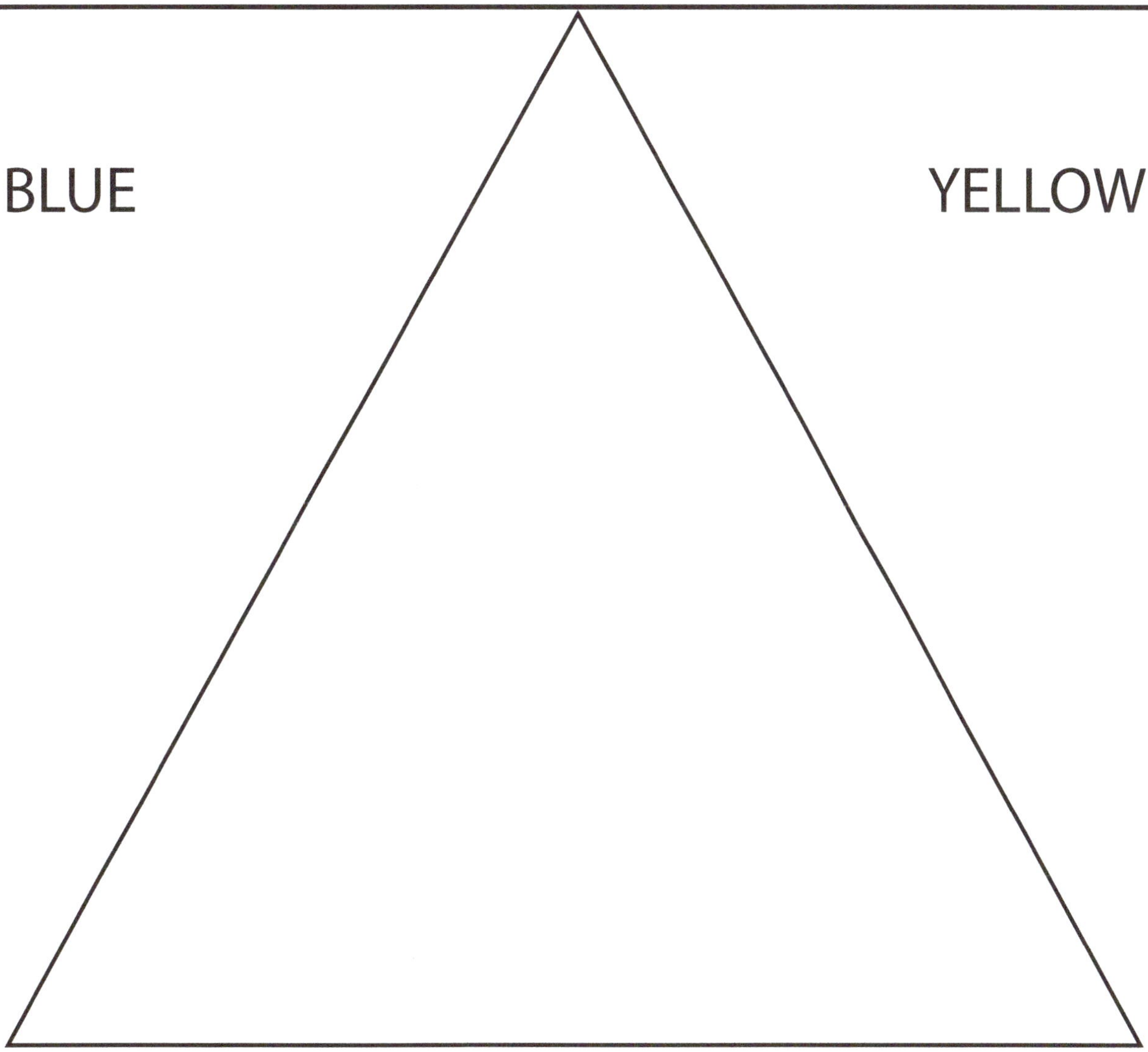

RED

TASTY TRIANGLE
STEMVESTIGATIONS: SHAPES

Re-shape your world. Use polygons to cover the image.

FLYING FIVES
STEMVESTIGATIONS: STAKE OUT THE PENTAGON

What lives on a pentagon of Earth? Draw what you see.

1. Use the pentagonal shape to stake out a piece of land.
2. Inside that pentagon is a lot of interesting stuff. Draw what you find.
3. Compare your discoveries with other kids in the class.

FLYING FIVES
STEMVESTIGATIONS: TOOLS

Tools make life easier. Make a tool for the classroom.

What activities happen in the classroom?

What tools do you use in the classroom?

What problems could you solve in the classroom?

What tool could you create that makes life easier?

On the back of the paper, draw your new tool!
Work with a team to design and build your tool.

FLYING FIVES
STEMVESTIGATIONS: PARACHUTES

Teacher Preparation or Accelerated Project

This plan was inspired by an ehow project.
http://www.ehow.com/how_8621360_make-pentagon-parachute.html
THESE PLANS ARE WILDLY NOT TO SCALE. USE YOUR VERY CREATIVE IMAGINATION.

MATERIALS
newspaper
protractor
pencil
string
kitchen garbage bag without handles or drawstrings

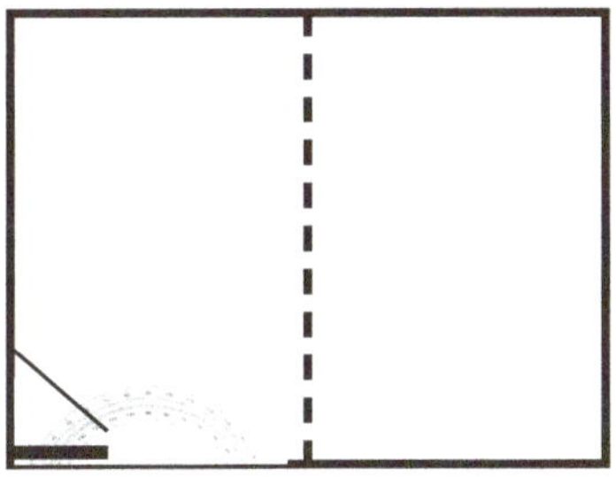

PREP YOUR NEWSPAPER TEMPLATE
1. Open a single sheet of newspaper on a flat work surface.
2. Lay your protractor along the bottom edge of the paper.
3. Create a horizontal 6-inch straight line along protractor's straight edge.
4. Place the center mark of the protractor on one end of the line.
5. Mark the newpaper at the 36-degree mark.
6. Repeat the process with the opposite end of the line.
7. Repeat the process on each of the newspaper's four sides.

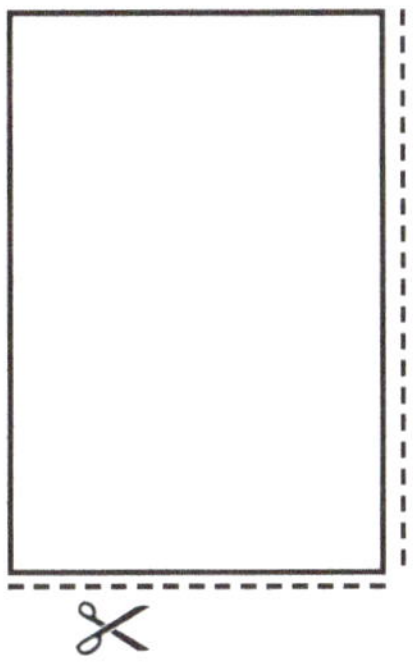

PREP YOUR GARBAGE BAG PARACHUTE
1. Place the paper pentagon pattern on the garbage bag.
2. Draw around the outside edge of the paper pattern.
3. Remove the pattern and cut the pentagon shape from the bag.
4. Arrange and flatten a garbage bag on the floor, table, counter, or other flat surface.
5. Open the bag by cutting down the side and across the bottom of the bag.

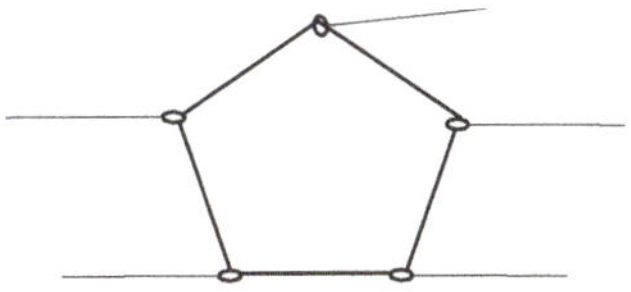

ATTACH YOUR FISHING LINE
1. Cut five lengths of fishing line approximately 12 inches long.
2. Tie one end of the fishing line into a knot around each of the five corners.
3. Tie the opposite ends of the fishing line to anchor weight.

FLYING FIVES
STEMVESTIGATIONS: STORY IN 5 BOXES

Create a comic strip story about your invention in five parts.

1. Comic strips tell stories in a graphic manner. Most of the story is told with pictures. The first comic was called Katzenjammer Kids.
   Introduce your invention. Explain why it is useful. Describe a way to use it. Give an example.

TITLE

FLYING FIVES
STEMVESTIGATIONS: 5-SIDED SHAPES

Walk The Pentagon. Test its angles with a pattern.

The Pentagon is one of America's best known buildings. Its design was chosen to fit an unusually shaped piece of land. When the site changed, architects and engineers kept the pentagonal shape of the building for its new location. Each edge of the Pentagon measures 921 feet. Walk 921feet. Walk it 5 times to simulate walking the perimeter of this huge building.

RHOMBI'S ADVENTURES IN 2D

MindBugs in Measurement

MODULE 1: SECOND GRADE

MindBugs in Measurement:

Read Across the Curriculum

Inch by Inch

According to encyclopedia.com, the inchworm is actually the "larvae of moths of the family Geometridae, a large…group with over 1,200 species indigenous to North America." These North American baby moths have become famous as characters in children's stories over the past 200 years.

Inchworms get their unusual name from the length of the body fully stretched. They really ARE about one inch long. According to the Collaborative International dictionary of English, "inchworms" are also called "measuring worms, spooners and loops." These names refer to the looping motion the larvae makes as it travels across the oak tree gobbling up leaves. The hungry inchworms first bring the rear end of the body forward, forming a loop, then move the front part of the body.

An inchworm's smooth, hairless body usually measures about 1 in. or 2.5 cm long. They are green, brown or black and may have lumpy projections that mimic tree twigs. If you scare an inchworm, it may stand still and tall (as tall as an inchworm can get) making them look even more like waving oak twigs.

MindBugs in Measurement:

Read Across the Curriculum

Think about Time

Why are there 60 seconds in a minute, 60 minutes in an hour, 24 hours in day or 24 time zones?

Think About History

Ancient Babylonians used base 60. They counted all the way to 59 before bundling 60 items. This base 60 system is still used in maps and clocks. When telling time, you count 1 minute to 59 minutes before bundling 60 minutes into 1 hour.

Time zones as we know them did not exist until the 1840s. 25 countries were involved in the decision to place the Prime Meridian so that it passes through Greenwich, England. In 1884, a vote took place in which 22 countries voted yes; 1 no; and 2 did not vote at all. More countries voted yes than no.

Think About Numbers

Units for measuring time help us organize our lives. As people tried to make smaller and more manageable units to describe and quantify the passing of time, the light and dark that defines each day provided a natural starting point.

The Earth rotates every period of light and dark (day and night). The Earth turns about its axis 360 degrees, or a full circle, in each rotation. The distance around the Earth has been broken into 24 even parts of 15 degrees each. This makes 24 time zones. Each time zone is measured with longitude lines on a map starting at 0 degrees at Greenwich, England.

MindBugs in Measurement:

Read Across the Curriculum

USING THE DIGITAL SCALE

IN THE BLINK OF AN EYE

A scale is a tool used to measure the amount of something. When an object is placed on the scale, the display screen will show you how much that object weighs.

PREPARE THE DIGITAL SCALE

Place the scale on the table. Press the (ON/TARE) switch. The numbers "8888" will appear. Wait for the display to read "0" before the scale is ready to use. Place the object on the weighing tray.

FIND THE WEIGHT OF AN OBJECT

Decide the units you need to record. Press "units" to choose grams, ounces, pounds (lbs), or pounds and ounces.

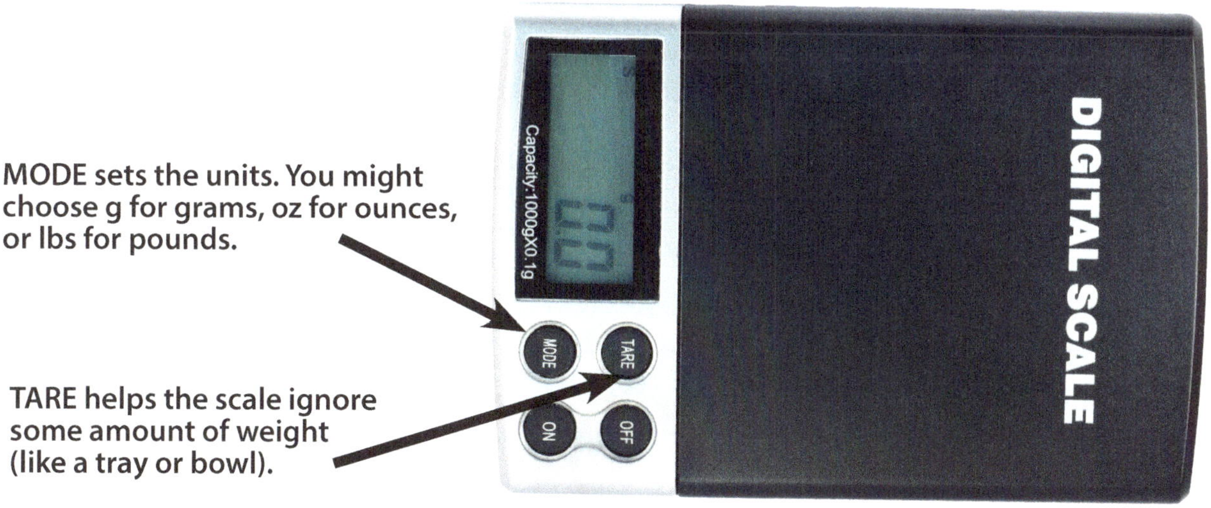

MindBugs in Measurement:

Read Across the Curriculum

Read about Coordinate Planes

SCIENCE FROM A SICKBED

Descartes is given credit for developing the system of coordinate points. A story tells of Descartes lying in bed during an illness, watching a fly move on the ceiling. He realized that he could describe the fly's location at any time using lines "drawn" together from two sides of the ceiling. It was this insight that supposedly led him to develop the grid or coordinate system we still use today.

WHAT'S THE POINT?

A coordinate point is usually written in parentheses and describes the x and y coordinates of a point. (2, 5) on a graph might tell the story of 2 rabbits that each ate 5 carrots. (2,7) on the same graph might tell us that 2 rabbits ate 7 carrots.

To plot a point, move across the x axis to 2, then up the y axis to 5. Where the lines meet, you'll place a dot.

Reading stories on a graph helps scientists, mathematicians, engineers, and statisticians understand the same story even if the storyteller is not present.

FINDING COORDINATE PLANES

Look up. Your classroom ceiling is a perfect coordinate plane. The ceiling tiles form a wonderful grid for plotting x and y. Hang a toy from the ceiling, label your dependent and independent variables, and then ask friends or classmates to tell you how to reach the item using Descartes' coordinate system!

MindBugs in Measurement:

Read Across the Curriculum

Using the Stopwatch

IN THE BLINK OF AN EYE

Can you use a stopwatch to time an eye blink? Try it. From measurements taken over time, we know that an eye blink takes place in a little over 15 hundredths of a second. That means you've cut a single second into 100 parts. The eye blink is just 15 of those 100 parts. Basically, a blink happens This means that if your reaction time is greater than 0.15 seconds, the blink is over before you can push the buttons.

Because people blink, the average person spends about 23 minutes in darkness ever day. This time in darkness may be why the brain has learned to "fill in the blank spaces" for us on occasion. It may be why "eye witness" testimony is not considered the best source of evidence. It may explain why several witnesses to the same event will have "seen" different things.

SLOW DOWN IN THE CAR

According to the book called Newton Rules Biology, messages move through your nervous system at about 160 feet per second. This speed translates to approximately 100 miles per hour, which explains why driving at high speeds is so dangerous. The speed on a highway is about 65 miles per hour. The speed in a school zone is usually 35 miles per hour.

If a driver is not anticipating the next move or looking ahead for other drivers in trouble, reaction speed can be slower than the speed of the car. Making the fast moves to avoid trouble may not be physically possible.

MindBugs in Measurement:

Read Across the Curriculum

Let Your Fingers Do the Measuring

For thousands of years, the human body has acted as a ruler for measuring. A finger's or hand's width and the length of your foot were all accepted ways to measure length. In the time of England's Round Table, when knights in armor fought battles on horseback, an inch was just the width of a man's thumb. How confusing that must have been! Compare your thumb to someone else's in the room, and imagine trying to build a castle using all the different measurements of a thumb's width.

THE IRON ULNA

Sometime after the Magna Charta was signed in the Thirteenth Century, King Edward I of England took a giant leap in measurement. King Edward knew that having everyone use the same measurement for an inch was important. He decreed that the inch should be one thirty-sixth of a yard. A permanent measuring stick made of iron served as a master standard yardstick for the entire kingdom. This master yardstick was called the "iron ulna", after the bone of the forearm.

AMERICA'S BRONZE BAR

Immigrants to the New World brought with them all kinds of different standards for measuring the inch and the yard. Finally, after some confusing decades, Americans decided that the inch and yard would be measured against the European metric system. America created a bronze bar that's yard measured 82 inches. The bar has been replaced at least once since it's first use in 1832.*

**http://www.cftech.com/BrainBank/OTHERREFERENCE/WEIGHTSandMEASURES*

SECOND GRADE MODULE 1: 2D SHAPES
VOCABULARY

Most vocabulary is defined within each story's context. The definitions on these pages address vocabulary outside the grade level and undefined within the story, itself.

astronomer
a person who is skilled in astronomy or who makes observations of celestial phenomena

autumn
also know as fall, this season lasts from the September equinox to the December solstice

builder
In this case, the term "builder" refers to a person that creates or develops a particular thing.

contractor
a manager and coordinator of all the tradespeople involved in a project of many phases

cylinder
a solid object with two identical flat ends that are circular or elliptical and one curved side. It has the same cross-section from one end to the other.

designer
a person who plans the form, look, or workings of something before its being made

engineer
a person who has scientific training and who designs and builds complicated products, machines, systems, or structures : a person who specializes in a branch of engineering

pi
the 16th letter of the Greek alphabet and the symbol π denoting the ratio of the circumference of a circle to its diameter

radius
length of a straight line from the center of a circle or sphere to any point on the outer edge

revolution
the action of moving around something in a path that is similar to a circle

rhombus
a shape with four sides of equal length and with four angles that are not always right angles

rotation
a complete turn around a central point

solution
something that solves a problem

telescope
arrangement of lenses or mirrors or both that gathers light and allows direct observation of distant objects.

RHOMBI'S ADVENTURES IN 2D
PRE AND POST ASSESSMENT PAGES

Give assessments prior to starting the projects so that you have a baseline for extension and reinforcement.

Give the same assessment after completing the projects to determine continuing areas of need.

Geometry
Numeracy
Place Value
Motor Skills
Equalities
Object Permanence (You will be surprised by this one.)

RECOGNIZE SHAPES

Color the square.	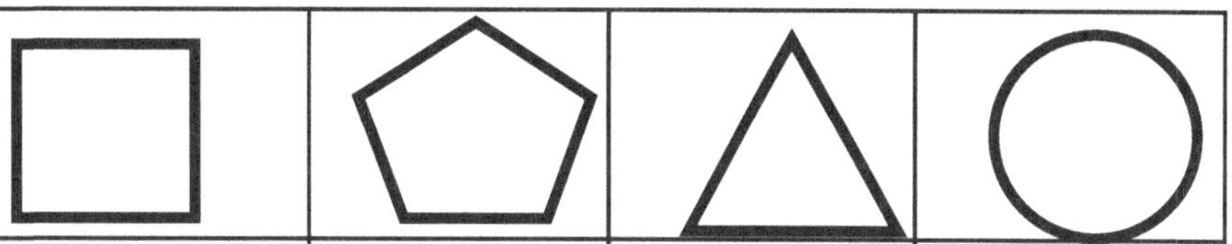			
Color the triangle.	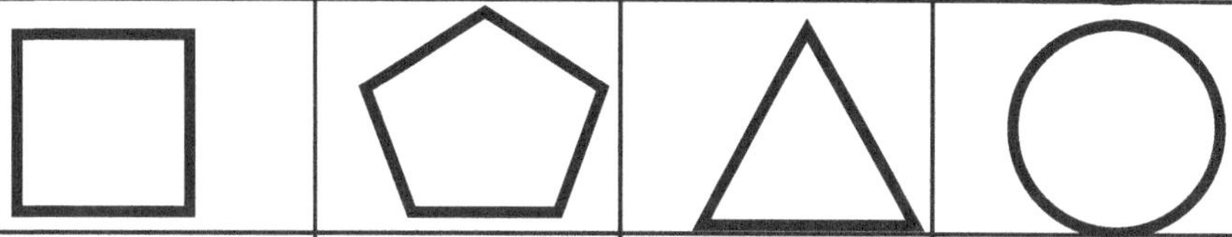			
Color the rectangle.	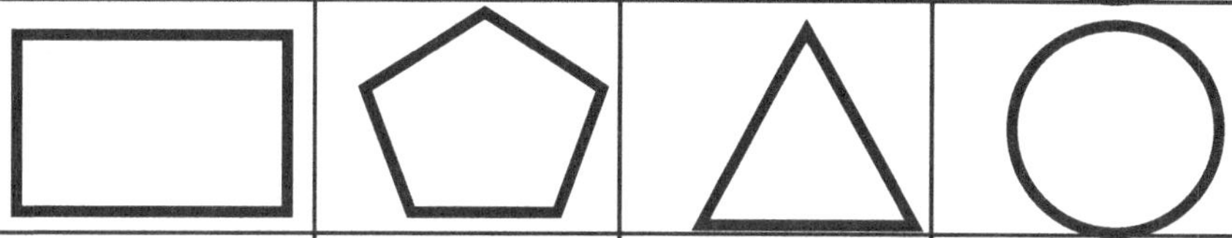			
Color the circle.	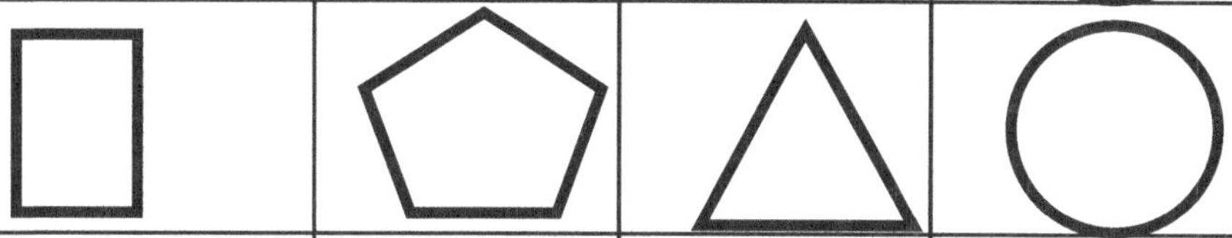			
Color the pentagon.	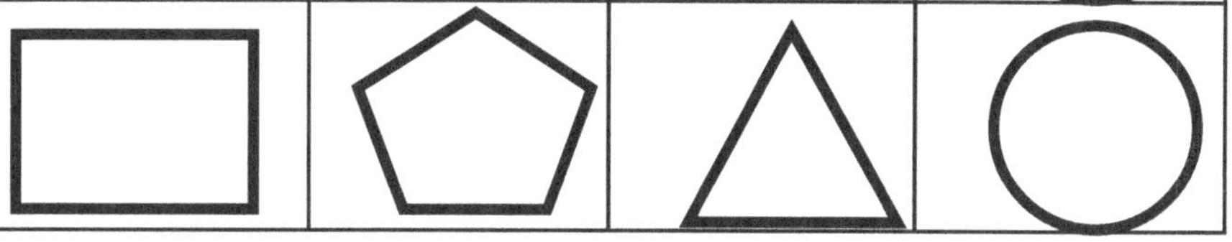			

REPRODUCE SHAPES

Draw a square.	
Draw a triangle.	
Draw a rectangle.	
Draw a circle.	
Draw a pentagon.	

MOVE FROM 2D TO 3D

Mark the shape you see on a cube.

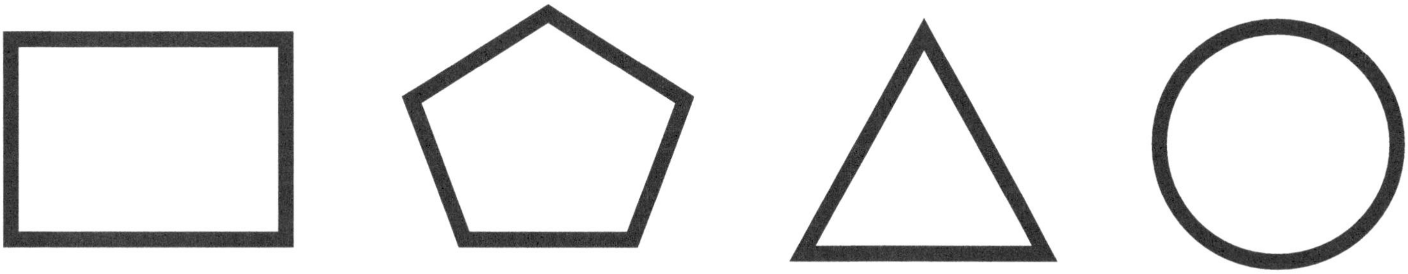

Mark the shape you see on a pyramid.

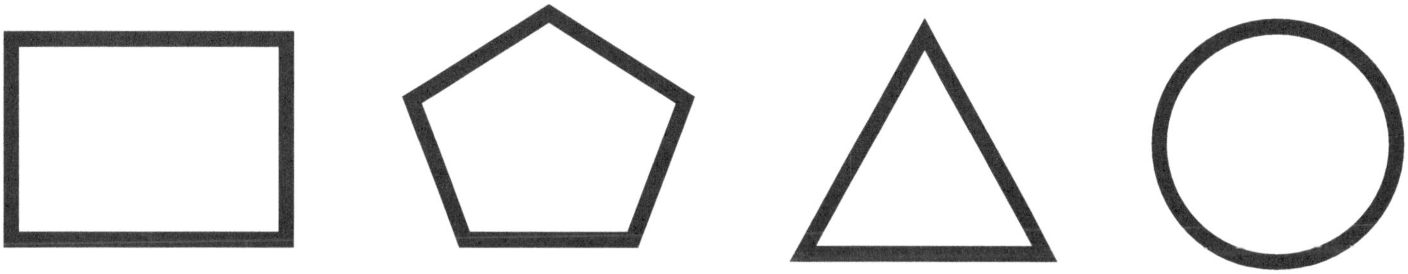

Mark the shape you see on a the end of a cylinder.

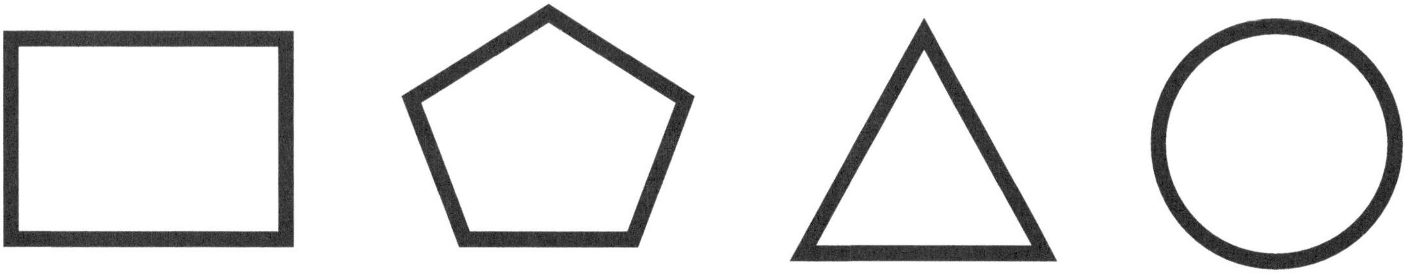

UNDERSTAND EQUALITY

Draw a line to break each shape into two equal parts.

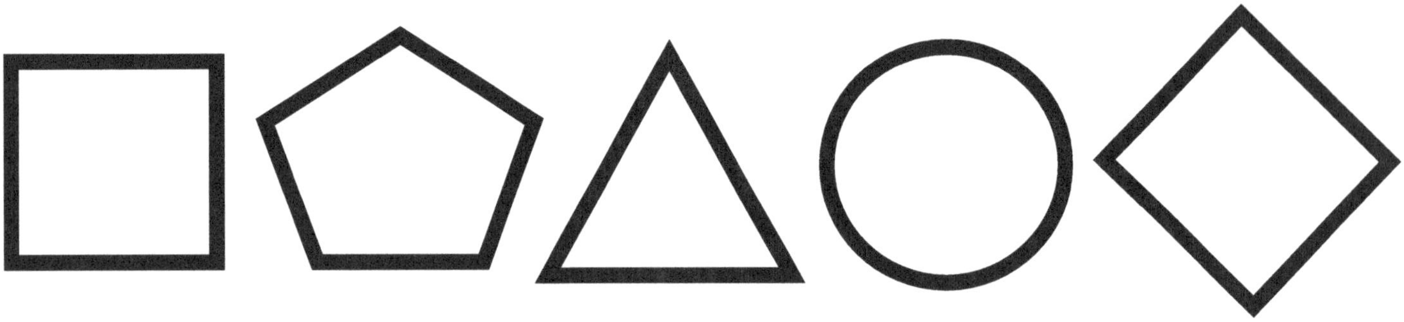

Draw the other half of each shape.

RECOGNIZE SHAPES IN THE WORLD

Glue the cut shapes to this page to make a picture.

DO NOT COPY FOR STUDENT
one-on-one with teacher

Show the child 2 pennies. Ask, "How many objects are in my hand?"

Child immediately says, "2." Has solid sense of abstract number 2...
Ready for discussions without visuals...

Child counts out loud and answers "2." Has basic understanding of abstract number 2...
Ready for handouts...

Child points to each object, counting out loud and answers "2."
Needs concrete objects to count while transitioning to abstract number sense...

Repeat actions for numbers 3 and 4.

Recognizing the sum of a pile of objects beyond 4 is almost impossible. Try it yourself. You will reflexively break the larger number into two or more lesser groups before combining to say the greater number.

All students will break numbers beyond 4 into smaller "piles." Note which students count by ones beyond 4 and which students automaticallly do some mental addition to give you the whole number.

ASSESS OBJECT PERMANENCE

DO NOT COPY FOR STUDENT
one-on-one with teacher

Show the child your hand as you fold 2 pennies into your closed fist. Reopen the same hand. Ask. How many objects are in my hand?"

Child immediately says, "2." Has solid sense of object permanence for 2...
Ready for discussions without visuals...

Child counts out loud and answers "2." Has basic understanding of object permance for 2.
Ready for handouts...

Child points to each object, counting out loud and answers "2."
Needs concrete objects to count while transitioning to object permanence...

Repeat actions for numbers 3 - 9.

Remember, in this assessment, your goal is to determine a student's grasp of object permanence NOT ability to count. The outcome is dependent on the student's ability to restate the cardinal number.

ASSESS NUMERACY & CARDINALITY

Print rubric for your records. Perform assessment as early as possible.

STUDENT NAME	1	2	3	4	5	6	7	8	9

NOTE: All students will need to break amounts into smaller groups in order to recognize the cardinal numbers beyond "4." The brain naturally grasps groups of 1,2,3,4. It's why phone numbers are broken into smaller groups.

RHOMBI'S ADVENTURES IN GEOMETRY
TEACHER'S PRINTABLE UNIT PLANNING PAGE

2D SHAPES: UNIT ______________________

STEMVESTIGATION	DATE	TIME	WHAT I'VE GOT	WHAT I NEED	PLANNING NOTES
SCIENCE					
TECHNOLOGY					
ENGINEERING					
ARTS					
MATH					
CHALLENGE					

RHOMBI'S ADVENTURES IN GEOMETRY
TEACHER'S PRINTABLE CHALLENGE PLANNING PAGE

2D SHAPES: UNIT ______________					
STEMVESTIGATION	DATE	TIME	WHAT I'VE GOT	WHAT I NEED	PLANNING NOTES
CHALLENGE:					

BRAINSTORM & HYPOTHESIZE	
DESIGN	
BUILD	
TEST	
ANALYZE DATA	

RHOMBI'S ADVENTURES IN 2D
NGSS CORRELATIONS: 2ND GRADE

NGSS K 2		***Module***	MODULE 1					
		STEMvestigations	ROUND	SQUARE	TRIANGLE	FIVES	MOVES	GIFTS
		Engineering Design Project	Cart	Tool	Bridge	Parachute	Move	
K-2	K-2.Engineering Design							
	K-2-ETS1-1	Ask questions, make observations, and gather information about a situation people want to change to define a simple problem that can be solved through the development of a new or improved object or tool.	x	x	x	x	x	x
	K-2-ETS1-2	Develop a simple sketch, drawing, or physical model to illustrate how the shape of an object helps it functionas needed to solve a given problem.	x	x	x	x	x	x
	K-2-ETS1-3	Analyze data from tests of two objects designed to solve the same problem to compare the strengths and	x	x	x	x	x	x
K	Forces and Interactions: Pushes and Pulls							
	K-PS2-1	Plan and conduct an investigation to compare the effects of different strengths or different directions of pushes and pulls on the motion of an object.	x		x	x	x	
	K-PS2-2	Analyze data to determine if a design solution works as intended to change the speed or direction of an object with a push or a pull.*	x			x	x	
K	Interdependent Relationships in Ecosystems: Animals, Plants, and Their Environment							
	K-LS1-1	Use observations to describe patterns of what plants and animals (including humans) need to survive					x	
	K-ESS2-2	Construct an argument supported by evidence for how plants and animals (including humans) can change the environment to meet their needs					x	
	K-ESS3-1	Use a model to represent the relationship between the needs of different plants or animals (including humans)		x	x		x	
	K-ESS3-3	Communicate solutions that will reduce the impact of humans on the land, water, air, and/or other living things in the local environment.		x	x		x	
K	Weather and Climate							
	K-PS3-1	Make observations to determine the effect of sunlight on Earth's surface.		x				
	K-PS3-2	Use tools and materials to design and build a structure that will reduce the warming effect of sunlight on an area.						
	K-ESS2-1	Use and share observations of local weather conditions to describe patterns over time.	x	x		x	x	
1	Waves: Light and Sound							
	1-PS4-1	Plan and conduct investigations to provide evidence that vibrating materials can make sound and that sound can make materials vibrate.						
	1-PS4-2	Make observations to construct an evidence-based account that objects can be seen only when illuminated.	x	x				
	1-PS4-3	Plan and conduct an investigation to determine the effect of placing objects made with different materials in the path of a beam of light.		x				
	1-PS4-4	Use tools and materials to design and build a device that uses light or sound to solve the problem of communicating over a distance.	x					

COMMON CORE
DO OR DON'T. THERE IS NO "I CAN."

A famous resident of Dagobah once said, "Do or do not. There is no try." Jedi Master Yoda may not sit with students in today's classrooms, but modern educational pedagogy would benefit from the concept. The use of "I CAN" statements is bothersome in that knowledge unshared does not contribute to community or society.

The "I can" is unnecessary. Let's help students focus on clear statements of ability. Action versus ability leaves little room for question. Goals are clear-cut, and students take pride in achievement. Visit the Ten80 Elementary wik site to download pages that help students take charge of their own learning goals. *(available December 2014)*

I count.
I read.
I name.
I write.
I participate.
I understand.

STEMvestigations
&
STEMathematics

ten80

Common Core Planning

Checklists for STEM Success

SECOND GRADE MATHEMATICS

SECOND GRADE MATH

Operations & Algebraic Thinking

2.OA.1.	**Use addition and subtraction within 100 to solve one- and two-step word problems involving situations of adding to, taking from, putting together, taking apart, and comparing, with unknowns in all positions, e.g., by using drawings and equations with a symbol for the unknown number to represent the problem.**	Represent and solve problems involving addition and subtraction.
2.OA.2.	Fluently add and subtract within 20 using mental strategies. By end of Grade 2, know from memory all sums of two one-digit numbers.	Add and subtract within 20.
2.OA.3. .	Determine whether a group of objects (up to 20) has an odd or even number of members, e.g., by pairing objects or counting them by 2s; write an equation to express an even number as a sum of two equal addends.	Work with equal groups of objects to gain foundations for multiplication.
2.OA.4	Use addition to find the total number of objects arranged in rectangular arrays with up to 5 rows and up to 5 columns; write an equation to express the total as a sum of equal addends.	
2.NBT.1.	Understand that the three digits of a three-digit number represent amounts of hundreds, tens, and ones; e.g., 706 equals 7 hundreds, 0 tens, and 6 ones. Understand the following as special cases: • 100 can be thought of as a bundle of ten tens — called a "hundred." • The numbers 100, 200, 300, 400, 500, 600, 700, 800, 900 refer to one, two, three, four, five, six, seven, eight, or nine hundreds (and 0 tens and 0 ones).	place value.

SECOND GRADE MATH

Operations & Algebraic Thinking

2.NBT.3.	Read and write numbers to 1000 using base-ten numerals, number names, and expanded form.	Understand place value.
2.NBT.4.	Compare two three-digit numbers based on meanings of the hundreds, tens, and ones digits, using >, =, and < symbols to record the results of comparisons.	
2.NBT.5.	Fluently add and subtract within 100 using strategies based on place value, properties of operations, and/or the relationship between addition and subtraction.	Use place value understanding and properties of operations to add and subtract.
2.NBT.6.	Add up to four two-digit numbers using strategies based on place value and properties of operations.	
2.NBT.7.	Add and subtract within 1000, using concrete models or drawings and strategies based on place value, properties of operations, and/or the relationship between addition and subtraction; relate the strategy to a written method. Understand that in adding or subtracting three-digit numbers, one adds or subtracts hundreds and hundreds, tens and tens, ones and ones; and sometimes it is necessary to compose or decompose tens or hundreds.	
2.NBT.8.	Mentally add 10 or 100 to a given number 100–900, and mentally subtract 10 or 100 from a given number 100–900.	
2.NBT.9.	Explain why addition and subtraction strategies work, using place value and the properties of operations.	

SECOND GRADE MATH

Measurement & Data

2.MD.1. .	**Measure the length of an object by selecting and using appropriate tools such as rulers, yardsticks, meter sticks, and measuring tapes.**	Measure and estimate lengths in standard units.
2.MD.2.	Measure the length of an object twice, using length units of different lengths for the two measurements; describe how the two measurements relate to the size of the unit chosen.	
2.MD.3.	Estimate lengths using units of inches, feet, centimeters, and meters.	
2.MD.4.	Measure to determine how much longer one object is than another, expressing the length difference in terms of a standard length unit.	
2.MD.5.	Use addition and subtraction within 100 to solve word problems involving lengths that are given in the same units, e.g., by using drawings (such as drawings of rulers) and equations with a symbol for the unknown number to represent the problem.	Relate addition and subtraction to length.
2.MD.6.	Represent whole numbers as lengths from 0 on a number line diagram with equally spaced points corresponding to the numbers 0, 1, 2, ..., and represent whole-number sums and differences within 100 on a number line diagram.	
2.MD.7.	Tell and write time from analog and digital clocks to the nearest five minutes, using a.m. and p.m.	Work with time and money.
2.MD.8.	Solve word problems involving dollar bills, quarters, dimes, nickels, and pennies, using $ and ¢ symbols appropriately. Example: If you have 2 dimes and 3 pennies, how many cents do you have?	
2.MD.9.	Generate measurement data by measuring lengths of several objects to the nearest whole unit, or by making repeated measurements of the same object. Show the measurements by making a line plot, where the horizontal scale is marked off in whole-number units.	Represent and interpret data.
2.MD.10.	Draw a picture graph and a bar graph (with single-unit scale) to represent a data set with up to four categories. Solve simple put-together, take-apart, and compare problems1 using information presented in a bar graph.	

SECOND GRADE MATH

Geometry

2.G.1.	**Recognize and draw shapes having specified attributes, such as a given number of angles or a given number of equal faces.1 Identify triangles, quadrilaterals, pentagons, hexagons, and cubes.**	Reason with shapes and their attributes.
2.G.2.	Partition a rectangle into rows and columns of same-size squares and count to find the total number of them.	
2.G.3.	Partition circles and rectangles into two, three, or four equal shares, describe the shares using the words halves, thirds, half of, a third of, etc., and describe the whole as two halves, three thirds, four fourths. Recognize that equal shares of identical wholes need not have the same shape.	

Distance and Speed

1st Grade Activity 3

Step # 1

What is your height in standard units of measure?

Hint: Standard units are those we use most often in America.

Step # 2

Name something 10 times longer.

16

CCSS English/Language Arts "I Can" Standards Reading: Informational Text Second Grade						
Indicator		Date Taught	Date Retaught	Date Reviewed	Date Assessed	Date Re-Assessed
Key Ideas and Details						
CCSS.ELA-LITERACY.RI.2.1	I ask and answer who,what,where,when,why and how questions to show that I understand nonfiction.					
CCSS.ELA-LITERACY.RI.2.2	I tell the main idea of a piece of nonfiction, including the focus of important paragraphs.					
CCSS.ELA-LITERACY.RI.2.3	I make connections between different events in history.					
CCSS.ELA-LITERACY.RI.2.3	I make connections between different science ideas.					
CCSS.ELA-LITERACY.RI.2.3						
Craft and Structure						
CCSS.ELA-LITERACY.RI.2.4	I figure out the meanings of words when I am studying a second grade topic.					
CCSS.ELA-LITERACY.RI.2.5	I understand and use all the helpful parts of nonfiction books to find important facts and details quickly.					
CCSS.ELA-LITERACY.RI.2.6	Ican tell the author's main purpose in nonfiction writing.					
Integration of Knowledge a nd Ideas						
CCSS.ELA-LITERACY.RI.2.7	I use diagrams and pictures to help me understand nonfiction.					
CCSS.ELA-LITERACY.RI.2.8	I describe how reasons suppport the points that an author is trying to make.					
CCSS.ELA-LITERACY.RI.2.9	I tell how the important points in two pieces of nonfiction about the same topic are the same and different.					
Range of Reading and Level of Text Complexity						
CCSS.ELA-LITERACY.RI.2.10	I read and understand second grade nonfiction by myself.					

VOCABULARY FOR GRADE 2
WORD WALL WORDS FOR MATH

Students should illustrate the words for a word wall. Just looking at the cute cards we all like to print and laminate is not particularly effective as a learning tool. While the student created cards take longer (and look significantly messier), the benefits are greater. tThe one who does all the work also does all the learning.

Addition	Value	Yard
Hundreds	Ruler	Meter
Skip count	Yardstick	Number Line
Expanded form	Meterstick	Rows
Standards form	Measuring tape	Halves
Number names	Foot	Thirds

Column	Cents	Angles
Line Plot	Dollar	Trapezoid
Picture Graph	Quarter (coin)	Half-circle
Bar Graph	Faces	Quarter-circle
A.M.	Vertices	Cube
P.M.	Pentagon	Quadrilateral
Hours in a Day Days in a Week	Octagon	Prism

FORCES & MOTION		SPACE
circular	machine	Earth
circular	objects	moon
fahrenheit	rotation	planet
force	revolution	Sun
gravity	vapor	astronomy

TOOLS		*a very few* STEM CAREERS
digital camera	balance **scale**	architect
Internet	digital **scale**	astronomer
magnifing glass	spring **scale**	engineer
3D printer	search engine	medical doctor, nurse, surgeon
ruler	stopwatch	programming
tape measure	telescope	scientist many fields

VOCABULARY FOR GRADE 2
WORD WALL WORDS FOR SCIENCE

DESIGN AND ENGINEERING		
construct construction	innovate	robot
data collect, represent, analyze	innovation	rover
design designer	machine	structure
estimate	mechanical	system
foundation	3D printer	weight

INQUIRY	LIFE SCIENCE	
chart	animal	habitat
graph	basic needs	insect
describe	cycle	pet
identify	Earth	plant
observe	forest	pollen
predict		seasons

RHOMBI'S ADVENTURES IN 2D
REFERENCE MATERIALS AND SUGGESTED READING

"Circle." Wikipedia, the Free Encyclopedia. Wikimedia Foundation, Inc, n.d. Web. 7 July 2014.

Dehaene, Stanislas. *Number Sense: How the Mind Creates Mathematics.* New York: Oxford UP, 1997. Print.

Devlin, Keith J. *The Math Gene: How Mathematical Thinking Evolved and Why Numbers Are Like Gossip.* New York: Basic Books, 2000. Print.

Ferris Bueller's Day Off. Dir. John Hughes. Perf. Matthew Broderick, Mia Sara, Alan Ruck, Jennifer Grey, and Jeffrey Jones. Paramount, 1986. Film.

Feynman, Richard P, and Jeffrey Robbins. *The Pleasure of Finding Things Out: The Best Short Works of Richard P. Feynman.* Cambridge: Perseus Books,1999. Print.

Furoy, Michael, Rosemary T. Wong, and Harry K. Wong. *The Effective Teacher.* Mountain View: Harry K. Wong Publications, Inc., 2009. Print.

Ifrah, Georges. *From One to Zero: A Universal History of Numbers.* New York:Viking, 1985. Print.

Lee, Joon S., and Herbert P. Ginsburg. "Preschool Teachers' Beliefs About Appropriate Early Literacy and Mathematics Education for Low Socio-Economic and Middle-Socioeconomic Status Children." Early Education and Development 18.1 (2007): 111-143. Print.

Ma, Liping. *Knowing and Teaching Elementary Mathematics: Teachers' Understanding of Fundamental Mathematics in China and the United States.* Hillsdale, NJ: Lawrence Erlbaum Associates, 1999. Print.

Mathemagic. Chicago: World Book—Childcraft International, Inc., 1980.

McLeish, John. *The Story of Numbers.* New York: Fawcett Columbine, 1994.

Next Generation Science Standards. N.p., n.d. Web. 7 July 2014.

Wassily Kandinsky – biography, paintings, books. N.p., n.d. Web. 7 July 2014.

"Inquiry Based Science: What Does It Look Like?" Connect Magazine. 1995:13.

Log onto the website to access additional lessons, activities, word wall printables, printable nets, videos, downloadable files, links, and other resources.

www.ten80elementary.com

Please contact us for more information about curriculum, programs, festivals, and professional development.

info@ten80elementary.com

To access the ArtofSTEM website and learn more about fundraising, or to support a school in your area, visit us.

ArtofSTEM.org

CPSIA information can be obtained at www.ICGtesting.com
Printed in the USA
LVOW01*0747081114

412604LV00004B/6/P